T0237701

"THE LINK"

From Before the Big-Bang to Brain and Beyond

Maurice Duffill

Order this book online at www.trafford.com
or email orders@trafford.com

Most Trafford titles are also available at major online book retailers.

© Copyright 2008 Maurice Duffill.
All rights reserved. No part of this publication may be reproduced, stored in a retrieval
system, or transmitted, in any form or by any means, electronic, mechanical, photocopying,
recording, or otherwise, without the written prior permission of the author.

Cover design by Maurice Duffi ll
(Utilizing Microsoft Clip-Art).

Note for Librarians: A cataloguing record for this book is available from Library
and Archives Canada at www.collectionscanada.ca/amicus/index-e.html

Print information available on the last page.

ISBN: 978-1-4251-7219-0 (sc)

Because of the dynamic nature of the Internet, any web addresses or links contained in
this book may have changed since publication and may no longer be valid. The views
expressed in this work are solely those of the author and do not necessarily reflect the
views of the publisher, and the publisher hereby disclaims any responsibility for them.

Any people depicted in stock imagery provided by Getty Images are models, and such images are
being used for illustrative purposes only.
Certain stock imagery © Getty Images.

Trafford rev. 02/06/2020

 www.trafford.com
North America & international
toll-free: 1 888 232 4444 (USA & Canada)
fax: 812 355 4082

This book is dedicated
to the great minds of past eras:

Socrates

Copernicus

Newton

Einstein

Heisenberg

Feynman

Those who taught us to *think!*

Also by Maurice Duffill

View of Life (autobiography)
End of Term (fiction)
Highland Haven (fiction)

Contents

Introduction – A Note to the Reader   7

PART 1 – The Warm-up – p7

Chap.1- Learning to Think   7
Chap.2 – Thinking of Science versus Religion   12
Chap.3 – Thinking of Thinking   20
Chap.4 – Thinking of Space and Time   27

PART 2 – The Hypothesis – p37

Chap.5 – Thinking of the Big-Bang   38
Chap.6 - Thinking of Evolution versus Divine Design   44
Chap.7 – Thinking of Waves   51
Chap.8 – Thinking of the Brain   60
Chap.9 – Thinking of the Common Denominator   73
Chap.10 – Thinking of Conscience   86
Chap.11 – Thinking of EGGS (or DEGGS)   97
Chap 12 – Thinking of Harmony   108
Chap.13 – Thinking of God   114

PART 3 – Review – p121

Chap.14 – Thinking Back   121
Chap.15 – Thinking of the Effects   129
Chap. 16 – Thinking of History versus Religion   160

PART 4 - Conclusion – p168

Chap. 17 – Thinking of the Explanation of Everything.   168

End Note   174

APPENDICES – p177

– The Puzzle and its Solution   177
– Electron Shell Structure per Maurice Duffill   178/179

Select Bibliography   180

INDEX   183

LIST OF DIAGRAMS

1 – THINKING   21
2 – Ripple Pattern   52
3 – Fundamental Sine-wave   54
4 – Harmonic Wave   54
5 – Fundamental + Harmonic   55
6 – Fundamental + 3.5   56
7 – 1st Shell   99
8 – 2nd Shell   99
9 - 3rd Shell   100
10 – The Grand Wave   105
11- Compound Characteristics   129
12 – Typical Basic Sub-Characteristics   130

Solution to Puzzle   177
Electron Shell Structure – P1   178
Electron Shell Structure – P2   179

Introduction

A Note to the Reader

This proposal introduces a concept which, I believe, ties together *everything* in human experience, both physical and metaphysical. This ranges from the space-time world of cosmology to the sub-atomic world of quantum physics. It embraces life itself, the mind, the spiritual world of human existence - even life after death. To identify a single basis for all these aspects of existence is a very tall order, hence we need to progress in a variety of directions before bringing the elements together. Only when the whole picture has been presented will the reader be in a true position to assess the value of this hypothesis. Accordingly, I suggest that the reader proceeds in a logical manner through the text; like a classical who-done-it, one must avoid reading the conclusion before reading the plot.

With the foregoing in mind the book is divided into four parts; the first part conditions the reader to the general idea of thinking afresh, the second part introduces the hypothesis being proposed, the third part examines how the common factor impinges on every aspect of life and death as well as being the seed of all matter, and the final part summarizes the 'explanation of everything'.

Maurice Duffill

Part 1 – The Warm-up

Chapter 1

Learning to Think

Physicists and mathematicians, for some time, have been searching for a unified theory which ties together all the principal laws of physics. Neuroscientists have been searching for an explanation of what causes the human brain to develop an awareness of self. Philosophers have joined in the search in an effort to explain consciousness. While theologians search for an acceptable explanation for the whereabouts of the soul, the public at large wonder how God can be 'in' us all and how much control He exerts upon us. Can it be that the final result of all of these searches will be found to have the same root? I think so. I believe *this* book is the Rosetta stone which shows the link.

I am neither a physicist nor a mathematician; I have not had formal religious training, and it happens that my command of the English language is not as well developed as one might wish. What then can *I* offer when, it transpires, that all of these attributes are needed to properly deal with the many questions considered by enquiring minds? What is life all about? Where do we go from here? What purpose has my life served? Is there a God? Is life pre-determined? Why am I who I am? Those of us who enjoy exercising the 'mentals' on philosophical matters will dream up many other questions as attempts are made to determine some answers to life's puzzles. What I have to offer is the output from an enquiring mind that has had training born out of the disembodied logic of electrical engineering. To properly understand the innermost workings of complex electrical circuitry requires that one's imagination has the ability to probe the simultaneous activities of a multitude of branches. Spending years in this field has exercised my brain in this very useful ability although my advanced years are now slowing that process.

Right from a very early age, I have been a thinking person; mentally querying the whys and wherefores of life as I saw it. I must admit that my thoughts have not necessarily been about that which I should have been thinking. Nevertheless, these mental sorties have revealed to me interesting issues in a wide variety of fields. This 'hobby' has become a habit - a potentially useful habit - which benefits from the application of method. In a later chapter, I will suggest how thinking can be organised - how different approaches can perhaps solve a problem or simply fill a void.

Many years ago, when I was a small child in England, I asked my primary school teacher a fundamental question and received a most unsatisfactory reply. The question asked was "Please miss, why don't Australians (who live on the opposite side of the world from the UK) fall off the globe?" "Oh that's because of gravity", was the reply. "What causes gravity, Miss?" was my next probing question. "That's created by the spin of the Earth", was her unreasonable explanation.

Thinking back to this incident, I have realised that it had great significance in my future development. Even at that early age - about 6 or 7 - I knew her answer was not correct; if you spin something, objects tend to fly off rather than be drawn in. From this episode a couple of lessons were learned: I learned not to always take statements as true - even from those in authority, without checking to ones own satisfaction [*I realised I had a thinking brain - even if she hadn't*]; Secondly, I learned that truth is of paramount importance - respect is lost for those who give false answers [*she had fobbed me off with the first thing that came into her head*].

The matter could have been dealt with, albeit without full explanation, if the teacher had admitted that she did not know the answer or had stated that the answer was too complex for me to understand, which it was - until comparatively recently. It was only while preparing this thesis that I truly thought I had the fundamental answer to the enigma of gravity. However, the positive outcome of this experience was the incidental lessons learned; these have had significant bearing on my thought processes ever since. I suppose, in a negative sort of manner, I should be indebted to her.

Many years ago, before our children were in their teens, as an excuse for them to delay going to bed, I would pose a thought provoking exercise for them to dwell upon while in dreamland. This was one mental problem I posed.

In this age of high speed flight, it is possible to fly faster than the apparent motion of the sun. Consequently, if one flies west, it is possible to arrive at a new location before the time of departure from the start point. For example, by

Concord, one could fly from London to New York, followed by San Francisco, Tokyo, Moscow, and back to London – a distance of just under 28,000 kms. in 14 hours –*ignoring details such as refuelling*. It takes the Sun 24 hours to track around the world, and so 10 hours have been gained. This being the case, one could dash around the world to carry out some task at home that had been forgotten such as buying flowers in time for one's wife's birthday.

Now in my seventies, my habit of thinking about what makes the world tick has amassed a great fund of material to draw upon. This library of experience stems from a diverse career in military flying and in engineering as well as living and working in various locations throughout the globe. To become a jet fighter pilot in the Royal Air Force it was necessary to be proficient in as many as 25 different subjects ranging from aerodynamics to law; from armaments to meteorology and from air-traffic control to medical matters. This required a great flexibility of mind. A 30 year career in engineering design and project development followed which took me from electrical switchgear through aircraft and super accurate measuring machine design to bulk materials handling which interfaced with a multitude of industrial processes. To create a successful design from a blank sheet requires considerable vision – imagination with a purpose. I mention these aspects of my life – not in an attempt to impress, but simply to indicate the background on which my thinking processes are based.

My life on this planet is concluding with a career in the hospitality industry - both public and private - spread across the United Kingdom, Malaysia and Australia. This requires an understanding of people. Having lived and worked in a city environment, in the country, the mountains, the jungle, and the Australian bush, as well as intimately experiencing a number of different cultures, I can fairly claim that any prejudices I may have had have been somewhat tempered.

It is because of this varied background that I was persuaded to write my autobiography, "View of Life", which led to the question; what have I learned from life? Trying to answer that focussed my many interests to the point where a 'theory of everything' evolved out of the soup of experiences and acquired knowledge.

Quite reasonably, we turn to specialists who are at the forefront in their particular fields to lead the way; to show what lies ahead in various directions. It is questionable, however, as to whether or not they are best qualified to take us to the ultimate – to the basis of everything. Inevitably, specialists become highly focussed with a tendency to develop a degree of tunnel vision. For example, while considering the dynamics of particles the physicist cannot allow himself to digress by thinking perhaps about the relationship of the soul to the body. Similarly, the neuroscientist, while theorising about axons and synapses is not concerning him/herself with the physics of the expanding universe. It is also unreasonable to expect a theologian to abandon his calling - backed by years of training, to seriously challenge the fundamental basis of his religion after learning of radical discoveries made by scientists.

On the other hand, the average person usually finds the intricacies of specialist subjects in the relativistic field of space-time, or the unpredictability of quantum physics, too great for the untrained mind to fully understand. The relevance of new discoveries and the resulting theories are often beyond one's comprehension. However, these shortcomings do not stop us 'lesser mortals' from *thinking* - from exercising the grey-matter. One can research what the specialists have to say so as to become aware of the general picture concerning a particular topic while keeping an open mind. In fact, to some degree it is an advantage to have a general awareness, supported by some detailed research; this way narrow reductionism can be avoided and a more holistic view taken.

Would you believe, the concept being presented within these pages which leads to a collective base for *everything*, embraces all material things from galaxies to globules or massive bodies of the cosmos to minuscule bodies of the atom. The transition from inanimate materials to living bodies leads to an understanding of the origins of life. The whole of life, - even 'life' after death - is found to have a common foundation. This concept gives insight into the development of human relationships, both in the short term and the long term. The process of assimilation, of learning right and wrong, is - at last - linked to the environment of upbringing. Art, science and religion are brought under the same umbrella.

Recognition of such an all embracing concept will, of course, require the acceptance of much flexibility of thought. It is not easy to remove any preconceived biases and to develop a truly open mind but this is absolutely necessary if the following hypothesis is to make sense. When about to engage in any significant physical activity, it is most desirable to perform warm-up exercises.

So it is with this mental exercise; conditioning of the thought processes is an important precursor. Let us then examine a variety of thoughts on different aspects of life before stretching the imagination to new limits.

Chapter 2

Thinking of Science versus Religion - historically

History suggests that, in order to explain matters not yet understood, various superstitions have developed which seem to give the cause of effects witnessed or experienced. A classic example of the need to dream up an explanation is the fear of thunder storms. It is quite understandable that children become disturbed during a storm – particularly at night. It doesn't help if the parent to whom he/she turns for comfort is also afraid and so, in ignorance, various tales are told to counter the mistaken belief that the gods are angry or the devil is flexing his fiery fangs. Only a true explanation will enable safety measures to be taken to give genuine comfort. Modern scientific investigations have revealed that there is a purely physical aspect of this phenomenon - albeit not yet fully understood. Another example: to me, it seems quite reasonable that early generations should worship the Sun; after all, it provides the daily light and the summer warmth - without which we would all be dead - or never even born. With the development of the physics leading to the atomic bomb, however, the inner workings of our nearest star (as well as others) are now basically understood, evaporating the need to imagine that it has any special supernatural powers.

Back in the 4th century BCE, the Greeks - who were then perhaps the leaders of early nation development - had what now seems a very strange notion of the heavens. Aristotle declared a concept of the universe which was accepted as a reasonable explanation of the visible sky above. The heavens were considered quite different to the Earth. In the heavens, the stars were supported on transparent crystal spheres - holding all bodies in a static relationship. In contrast, on Earth everything is subject to change giving an appearance quite different to the surrounding celestial orb. This theory generally held sway for nearly 2000 years; the hierarchy of the church favouring the idea of heaven having different characteristics to that of Earth.

Much to the consternation of religious bodies, in 1572 the Danish astronomer Tycho Brahe observed a 'new' bright star in the constellation of Cassiopeia. Actually this was an old star exploding – what we now refer to as a supernova. The heavens had changed; they were no longer the stable supernatural structures enveloping and nurturing the whole of mankind. Aristotle's fanciful idea was finally killed off by the arrival of a comet in 1577 which generated intense astronomical study; this further demonstrated that the heavenly sphere was not unchanging. Maybe this was the start of the rot; the relationship between science and the Church would never be the same. Who should the populous believe; The Church or the scientists?

On the other hand, alchemy gave science a bad name; by mixing magic and mysticism with chemistry people were misled. Right up to the latter part of the 18th century, chemical trickery served to obscure the true nature of science; giving ammunition to the faithful in their argument against the, so called, blasphemers. This, no doubt, caused people to largely ignore the progress being made by the great cosmologists, physicists and mathematicians from the 15th century onwards to the more enlightened recent centuries...

Similarly, in more recent times, religious doctrines have fought against Darwin's theory of evolution. Narrow-mindedness or religious dogma has caused some factions to deny the possibility of its truth because the theory of natural-selection contravened the literal word of the Bible. Some hard-liners still take this view, creating conflict between creationists and evolutionists. On the other hand, Darwin's "Origin of the Species" has been deliberately misquoted and used to further the aims of puritanical factions. This has been particularly so in America where Herbert Spencer developed what has become known as Social Darwinism. In this doctrine it is believed that the rich are destined to become the super race at the expense of the less fortunate who will be cast aside as inferior. Similarly, German policy in the 1914-18 war was influenced by the doctrine of Monism, based on Haeckel's "History of Creation", in which a ranking of 12 human species proclaimed Germans as being of ultimate intelligence.

Even today, we have great controversy regarding the use of genes; whereas they can be used to develop improved strains of plants, they can also identify inferiorities, or imperfections, in individual humans. More importantly, this sphere of scientific development has evolved into the ability to clone living beings. The implications of such development are highly complex from the scientific point of view but also have raised powerful ethical arguments, but these are often based on misconceptions. One needs to be informed while keeping a relatively open mind, in order to make proper judgements.

In recent times the previously discredited evolutionary theory propounded by Jean-Baptiste de Lamarck has regained credibility. In this theory, Lamarck claimed that characteristics acquired in a lifetime could be passed on to the next generation. Much later in this work we will discuss how this might - with reservations – have some relevance although Darwin rejected it.

It is not too surprising that some of the less informed have that underlying distrust in the great strides currently being achieved in the scientific world; to them, scientific proclamations appear to be all mumbo-jumbo. Modern day media constantly adds to the confusion by frequently misquoting the details of new developments, generating false hopes and ideas. This is particularly so in the medical field, with new cancer cures being announced on a more or less weekly basis.

Before the birth of Christ, religions had developed which provided cohesive doctrines to supersede the fractious notions of the uneducated masses. Maybe as far back as 2000 BC Druids had developed a doctrine which lasted into the Christian era. Judaism, based on the declaration of Moses as contained in The Torah, has lasted from over 1200 years before Christ to the present day. There not being printed words in ancient times, travellers spread the word of the creation and how mankind has reportedly developed, by stories handed down from generation to generation throughout the whole globe. Inevitably these have become distorted in the re-telling but they were never based on sound knowledge in the first place.

It is quite understandable that with the birth of science - the search for physical truth, the new men-of-science dangerously undermined the authority of religious leaders by implying that the men-of-God were less than truthful. This came to a head in the 16th century with the work of Copernicus, Galileo, and Newton, followed later by Einstein and others with theories - subsequently proven, which appeared to further challenge the then accepted explanations of natural phenomena offered by the Church.

The odd thing about this quartet is that they were all religious men - Nicolas Copernicus becoming the Canon of Frauenburg Cathedral. It was he that really set the ball rolling with his declaration that the Earth was *not* the centre of the universe as was originally envisaged. He showed that we lived on a planet which, amongst others, orbited the Sun - just one of many stars. We now know that not only is the

14

Sun one of millions of stars in just one galaxy - the Milky Way - but it, in turn, is one of millions of galaxies set in clusters which, along with others, are distributed throughout the universe.

The mind boggles at the thought of these vast numbers and the expansive distances between the multitudes of bodies in this great cosmos. However, as modern cosmologists study this fascinating universe with the aid of space probes and massive telescopes - both visual and radio - the general public is treated to dramatic film sequences of exploration amongst the heavenly bodies. More and more strange bodies are discovered and introduced to the ever more aware armchair viewer; pulsars, quasars and black holes, to name but three.

It is the realisation of the existence of this multitude of bodies that triggers many questions. Where are the absolute boundaries? How did it all begin? How did mankind develop from the inanimate elements of cosmic dust? How and when will it come to an end? It also raises the possibility that we are not alone, that other intelligent beings have developed elsewhere - even perhaps before us. It is these very reasonable questions that appear to challenge the traditional teachings of the Church - regardless of particular religious doctrines.

Eventually the ultimate questions arise: Is there a God? If so, what control does He exercise over our lives? How can He be 'in' us all?

It may be considered profane for me to raise such questions but I am certainly not the first to recognise that there must be a connection between science and the ultimate truth. Francis Bacon argued, in 1247 AD that good science is a handmaid to religion - not a challenge to it. Even earlier, the French philosopher, William of Conches, argued that God operates through natural forces. As I understand matters, science is the investigation of these natural forces and, therefore, is the exploration leading to a greater understanding of our creation - and our Creator.

General religious teachings can be likened to the statistics expounded in the popular press. Misapplication of statistical figures can frequently be the basis of totally misleading statements. For instance: the development of the mobile phone in Australia is hampered by the vast open spaces of the outback and desert. Recent developments in the communications field have led providers to claim that the new system reaches 98% of the population - implying that the signal reaches most parts of the country. This is totally misleading since the majority of the population live in the major cities which are situated around the coastal belt. Move inland and coverage disappears. Where I live mobile phones do not function because we are situated in an isolated valley, shut off from signal towers. It is probable that coverage here will never be provided for the ordinary cell-phone because my wife and I are currently the only permanent residents – the economics would make the prospect unreasonable; but the fact is this same valley is a popular target of campers and caravaners who frequently have the need to communicate.

Similarly, selecting biblical quotations out of context can give apparent authority to conflicting views. The Bible tells us that, in conflict, we should 'turn the other cheek' but elsewhere it is suggested that we should 'take an eye for an eye'. To understand what lies behind these two statements requires an understanding of the context. Making such dogmatic statements in support of one's beliefs is not at all helpful. We need to believe in *something* but we should also learn to *understand* in order to rationalise our beliefs. True understanding can only come from fundamental truth. Einstein summed it up when he stated "Science without religion is lame. Religion without science is blind."

Modern generations have a tremendous advantage over those of 2000 years ago; so much true knowledge has been gained. Through the development of the mass produced book and latterly, electronic communication, this knowledge has largely been passed on to the ordinary man in the street. We live in very exciting times, not only do we know the world is round and that it orbits the Sun, we are generally aware of the whole structure of the universe and its component parts.

It is quite understandable that ancient man invented gods of this and that, as a substitute for true knowledge; 'somebody' had to be the cause of things happening. Scientific discoveries are progressively showing what causes what to happen. We are now much better informed and, generally, much better educated so we should apply our new found attributes to re-appraise the situation. The object of this book is to indicate the thread which binds everything together – giving the ultimate explanation of what makes the universe tick.

Traditional church schools can be a bit of a worry; the church focuses on strengthening belief whereas schools should focus on discovering the truth. I hope that modern school governors are sufficiently enlightened to allow science to clear away the superfluous parables so that one's beliefs have a firm foundation.

It has been said that science, particularly Mathematics, has elegance; if it looks right it probably is right. If this seems like a biased observation, I ask you to consider Isaac Asimov's Tree of Science. He observes that the beauty of a tree is commonly attributed to the leaves since they are what draw the attention, particularly in the spring (or autumn). However, they are only the leading edge; marking the progress of the twigs supported by the branches. These, in turn, are held in place by the trunk which relies completely on the foundation provided by the roots.

Similarly, new scientific theories draw attention as the leading edge of current thinking but these theories are not necessarily correct. One should not judge on the exterior view alone since all new work - as with the tree in spring - is based on the underlying structure long since established. The currently observable face of science is the result of generations of work. If the end result is subsequently proved faulty, it is pruned off - bringing new life to that branch. Only the tried and true development of knowledge remains, forming a sound basis for future development.

It is, therefore, the basic form which has elegance; if the structure is right there will be no need to introduce superfluous detail. A leafless deciduous tree in the winter can still display a good form and structure. This even applies to fashion, either that of clothing or perhaps architecture. If the basic form is shaped appropriately, with originality of style and innovative features, to add decorative detail does not improve the fundamental purpose and may even detract from the beauty of the whole. No amount of false decoration will make an unsound design satisfactory.

In the same manner, any religious calling will gain strength and purpose if it is based on a firm foundation of truth rather than the flimsiness of myth and legend. It is realised that, before the advent of true science, stories had to be devised to explain observed phenomena. This was particularly so bearing in mind the inability of the masses to read and write and the lack of availability of the printed word. In modern times, with the much higher level of general education, we must – as individuals, be prepared to upgrade our basic thinking in keeping with the advancing horizon of established knowledge. A proper appreciation of new discoveries must lead to a deeper understanding to the betterment of the whole of mankind. After all, science is simply the discovery of what is already a fact of nature.

Regrettably, we are all aware of the misuse of knowledge – development of the atomic bomb might well be considered the classic case - but we should not confuse these errors of man's responses with the true progress of knowledge. Even the principles of this awesome weapon have now been applied to the betterment of mankind. Our behaviour is a humanistic problem outside of the realm of science. History has shown that man has fought man with ever more effective weapons long before the birth of science. Military needs have, however, been the impetus for much technological development which has accelerated mankind's evolvement.

I have followed with great interest the deliberations of the cosmologists as they explore the expansiveness of the universe. I have also - with great difficulty - attempted to grasp an understanding of the infinitesimal world of quantum physics, where matter is investigated that is so small it appears to exist only as a mathematical probability. One may, with some thought, be able to visualise the vastness of the universe - we can, at least see some of it. It is very much more difficult to develop a mental picture of what is happening in the world of quarks, the sub particles of the atomic nucleus, which in turn, are minute constituents of the molecule - the building block of everything, including us.

Research into the massive components of the universe around us, or the minuscule domain within us, does not seem to lead to any definite guidance towards answering the fundamental questions referred to earlier. Considering the whole rather than the reductionist detail, the general conclusion I have come to is that human experience has provided overwhelming evidence that there *is* a God; but not I hasten to add, as generally portrayed in the accepted religious tenets. However, those readers with a strong religious belief need not fear the outcome of this proposition; it does not seek to destroy one's fundamental belief but should lead to a greater understanding of the basis of everything, including spiritual awareness.

As will be seen in due course, God's All-Enveloping-Influence is not challenged. In fact the mystery of how God is *in* us all is explained. Personally, I could never understand what was meant by the biblical declaration that "man is made in His image". I now know how this can be true - not just a bald statement to be accepted as a religious doctrine. The resolution of this enigma will be revealed as we progress.

What then is this? Having highlighted many potentially very deep questions, I have already declared a fundamental conclusion; we appear to be starting at the

18

end. This is quite deliberate so that this basic statement of my position can be grasped as a thread throughout what follows. I have not arrived at such a conclusion without travelling a long road of thoughts based on a quite diverse existence on this planet. Many lessons have been learned in an assortment of ethnic environments. With this grounding, it is my hope that we can take a potentially interesting journey together, along this winding road, while exploring varied thought processes.

Although it is the task of scientists to seek the facts regarding the processes of nature, we should not consider truth to be their sole domain; people have their own thoughts and beliefs which need to be guided. I hope to show that to believe in God need not be to believe in a rather vague supernatural power. The point is, in order to have meaningful thoughts one should develop the habit of thinking. Let's learn to think things through at least to the satisfaction of one's self.

Some 400 years before the birth of Christ, Socrates spent the latter part of his life propounding the importance of thinking for oneself. He realised that the opinion of the majority did not necessarily mean that their view was correct; we tend to 'go with the flow' without putting the alternatives to the test of logic. Without implying superiority of his intelligence, he reasoned that the majority of the populace did not have the mental ability to 'think things through'. I can't help but agree with this opinion and go further by stating that this shows the flaw of the democratic process. Since the vote of the couldn't-care-less or the ill-informed person carries just as much weight during an election as the politically aware person, it does not seem reasonable to make it mandatory to cast a vote as it is in Australia; it causes the decision to fall to the pen of the followers (the majority) rather than the views of the thinkers (the minority).

We should all learn to think for ourselves by mentally challenging generally accepted views before blindly following ill considered doctrines.

Chapter 3

Thinking of Thinking

It is not everyone who bothers to think about matters encountered in everyday life; some see little merit in concerning themselves with abstract affairs. However, if you *are* someone who *does* give consideration to the 'whys and wherefores', it helps if some suitable methodology is adopted to deal with any subject of interest.

As an engineering designer, I found it was essential to think objectively of all the aspects associated with each and every project. Starting with a blank sheet of paper on a drawing board can be rather daunting; so many questions are raised. Where do we start? What is needed? How are we going to achieve the objective? What hidden hazards lie beneath the surface? Of course, other questions such as costs, or the available time-frame also have great bearing on the final outcome.

With so many questions raised before even getting started, one could be excused if panic set in or a nervous breakdown resulted. On the other hand, great excitement can be generated at the prospect of providing a satisfactory solution to a difficult problem. The thrill of creating something new, never seen before, can certainly make the adrenaline rush. To achieve a satisfactory result, however, it is necessary to organize one's mind

The ability to think things through does not come without practice. It is a habit which develops over a lengthy period of time; a habit born out of challenging oneself to find the answers to all manner of questions. We've already seen how, even as a young child, I had the impudence to challenge authority in an attempt to uncover the truth

As an engineering designer it is not surprising that I like to think in pictures, so let us see how that concept can be applied to developing a method for organized thinking.

On the following page is my mental picture, in diagrammatic form, of the major aspects of the process of thinking about a problem or a plan of action. There are four sectors to the diagram, each leading to a different area of mental activity. I suggest that practising these approaches to organizing the mind can be helpful in many day-to-day situations.

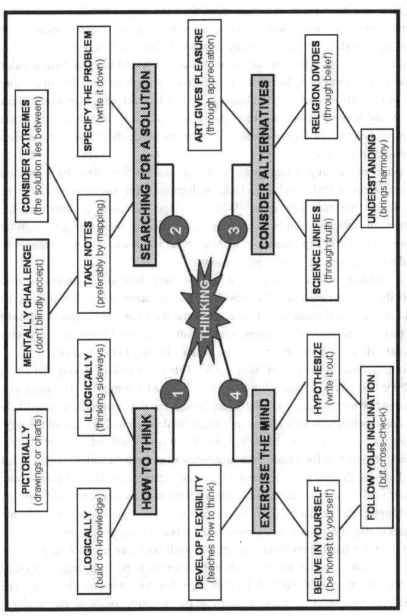

Diagram 1 - Thinking

Let's take time to examine the main elements - the way I visualize the process. Starting at number one, we see that three ways are suggested as basic methods of organizing thoughts. To think logically appears obvious but it is not everyone that can; scatterbrain characters are met in all walks of life and without them a lot of sparkle would probably be lost. Even the relatively straightforward task of moving house or travelling extensively on business or holiday without a plan would court disaster. By writing down important dates, lists of needs and costing, a basic framework is made for logical thinking; one begins to see the extent of the project and possible problems.

In professional project management, the process is formalized by a method known as Critical Path Analysis which, in diagram form, shows *what* has to be done *when*. One can then apply 'What if?' type questions to determine the likelihood of difficulties. In general terms, it is logical to think things through on the basis of personal experience - to build on what is known - and seek sources to extend one's knowledge of what is not known.

If a problem seems intractable, it can be very helpful to discuss it with somebody else. The mere act of composing an explanation of the problem sets the details into a logical sequence which can - sometimes - reveal the answer. I have done this on a number of occasions; spelling out a technical problem to my non-technical wife has exposed the root of the matter - leading to the solution.

If time permits, it can pay to 'sleep on it'. This is not just a delaying tactic or a vain hope that you will have more time to think about it tomorrow. If, just as you are settling down to sleep, you briefly run through the day's work, it is quite likely that your subconscious will review any snags while the mind is clear of the day-time clutter. In the morning - EUREKA, a solution comes without apparent effort.

While responsible for a team of draughtsmen working on a variety of projects, I tended to have the bad habit of being a little late to the office. Most of them would not start work until I arrived because, two or three times a week, I would start the day changing the design to overcome a problem that had been resolved in my thoughts during the night. It was almost as if I was receiving divine guidance.

On the other hand, a problem may present itself such that no solution comes to mind. How can you think logically when there seems to be no avenue to pursue? Then we can try another approach by thinking sideways - thinking illogically. The classic case presented to trainees aspiring to executive ranks is the practical problem of crossing a river without a bridge. There may be bits and pieces of paraphernalia lying about, none of which are large enough to span the raging waters. Play with ideas, list anything that comes to mind - whether reasonable or not - then 'kick' them about.

At least, the mind is focussed on the problem. It just may be that some fleeting crazy notion has an element of sense which, combined with other thoughts can provide the key. An example of this technique being successful was when my thoughts were exchanged aloud with a colleague, resulting in a concept for a machine to manufacture microscopic steel balls for sub-miniature ball bearings - something the professionals in that field had failed to achieve.

The idea of representing notions as pictures is certainly valid; relationships between elements reveal much about the structure of a concept. Draughtsmen draw to deliver the designer's solution to a puzzle presented by a client. This very diagram we are discussing conveys so much with so few words; it conveys a structure and suggests avenues to explore without cluttering the mind with unnecessary words.

Artists paint or sculpt the feelings they have for a scene, for people, or for abstract ideas; their work conveys, visually, the thoughts of the artist. The popularity of television owes much to the readiness by which the viewers' minds absorb information from pictures. The personal computer was never going to gain the current level of acceptance while the screen presentation was limited to basic DOS displays. With the advent of Windows, or a similar type of array, non specialists found they could soon learn to explore the computer's potential.

Mathematics, the bane of many people's education, used to appear to relate to pictures only via graphs, vector diagrams and charts. In recent decades it has been shown that certain types of mathematical expressions are really the embodiment of fantastically elaborate, and very aesthetically pleasing, pictures known as fractals. Awareness of these pictures has only come about with the advent of the digital computer since millions of self-progressing calculations are required to generate the display. 'Chaos' is the general name given to this developing field of mathematics in which these types of calculation are finding value in many spheres of complex study ranging from population growth/decline to turbulent flow of fluids. It is, however, worth noting that nature has developed these complex relationships, not the mathematicians – they have only discovered they exist.

Nature has provided a direct link between art, science and mathematics; these fractal pictures are the spontaneous illustration of the details, of the details, of the details, ad-infinitum – just as there maybe fleas on the backs of fleas – on the backs of fleas.

On a much simpler level, I was recently given a puzzle to solve - simply as a fun challenge. I must admit to sending my thoughts around in circles trying, vainly to solve it until, finally I applied my own rules of sideways thinking and pictorial representation; then, hay presto, everything dropped into place.

It is such a good example of all three approaches to the basic thought techniques that I have repeated it here.

A LOGIC PUZZLE

The ages of Anne and Bill summate to 91 years.

Anne is three times the age that Bill was
when Anne was twice as old as Bill is now.

What are the ages of Anne and Bill now?

At first, it seems impossible to answer, but it isn't. Why not try it for yourself? I suggest you think in terms of graphical pictures. Somewhere later in the book I may help with my answer - just in case you get stuck.

So then, we tackle a problem logically or illogically as needs dictate - presenting the details pictorially either to self or to others but how do we actually search for the solution? Let's move on to leg 2 of the diagram.

Searching for a solution to a problem is the usual impetus for thinking; we need to know what to do. Whilst considering options, it helps to take notes of the pros and cons of the ideas which pass through the mind. It might be that element 'A' will not work with the first possible approach but could be of use with subsequent notions. If these notes are in diagrammatic form, arrows, side notes and other scribbles can help form a picture. Only with a good overview can one see the 'wood' and not just the 'trees'.

Mentally challenging accepted practices or procedures, even if subsequently found to be in accord with previous knowledge or experience, can lead to a fuller appreciation of the basis on which they are built. It might also lead to new roads being found.

Considering extremes is sometimes quite a useful tool. Let's take, as an example, a simple problem requiring a decision. Suppose a couple wish to purchase a new dining table. How large should it be - sufficient for just the two of them, or large enough to accommodate a family party of say twenty? No doubt one is too small while the other is too large. Obviously a compromise solution has to be sought, taking into consideration the size of their accommodation, the likely frequency of dinner parties and, of course, the cost.

Leg 3 of the diagram suggests that alternatives should be considered. I'm referring here to general thought trains - one's own philosophy of life. Whereas it is commendable to be totally focussed on a task, tunnel vision should be guarded against. We have all met narrow-minded people who seem unable to even consider the points of view of others. They may be staunch vegetarians who believe killing animals for food is cruel, or rigid environmentalists who are unprepared to compromise their ideals even for the betterment of others. I believe that everyone is better off sociologically if an open mind is maintained by all.

Presumably it isn't surprising that, having an engineering background, I take an interest in scientific discoveries. I have come to believe that these discoveries lead to the truth; the truth regarding natural occurrences, man's development, even the origin of the cosmos. It saddens me that this general observation is not universally accepted. Bigoted believers still hold on to ingrained dogma that does not allow for free thought. I believe that it is these mentally closed factions which cause so much conflict in the world. Engaging in collective communion should not cause a group to reject the views and preferences of others. The result of doing so is to divide communities sometimes to the extent of causing violence - even war. Understanding others' viewpoints should occupy much of our thoughts.

Then again our thoughts need not be confined to matters of fact or matters of belief. The various art forms; painting, sculpture, music and dance, and many others all provide much food for thought. One can be absorbed by a masterly painting or lost in the beautiful harmonies of fine music. Art becomes a great aside to the rigors of daily life. Soak it up - then let the mind wander amidst attractive thoughts to counter the nasties of the world.

The final leg of the diagram suggests that the mind should be actively exercised. Doing so develops the flexibility necessary to take full advantage of new experiences. By exercising the mind one can travel - without restriction - to wherever and whenever the inclination takes you. It can be a serious exploration of the 'whys', 'wherefores' and the 'what-ifs', or a fantasy journey into an improbable world. It is an activity that can be performed whenever we choose; there is no need ever to be bored or to feel alone while the intellect is able to pursue the subject of interest. It could also possibly result in a new idea, a step forward in the quest for knowledge.

As with most things, we see that being organized or at least having objectives and the method to achieve them, can result in greater efficiency; in this case, efficiency of thought.

With our minds conditioned to organising our thoughts let us think about space and time, the environment in which the Earth and all other bodies - including us - exist.

Chapter 4

Thinking of Space and Time

Since space embraces everything, perhaps the fundamental unit of measurement should be the cube. Nothing actually exists in one dimension (a piece of string has a cross sectional area), nor in two dimensions (even a sheet of paper has thickness). If we use the three dimensional cube as our basic unit, area and length become merely subdivisions of that fundamental. Visualising a cube gives us a mental picture of space. With the cube as our unit of space, we can then appreciate better the combination of space-time.

We are accustomed to seeing shapes, rectangles, triangles, circles, etc. on flat sheets of paper or whiteboards and, nowadays, on computer and television screens - always in two dimensions - but these are unnatural. A rectangle or a triangle drawn on the Earth's surface is actually curvilinear - due to the spherical form of the globe - rather than rectilinear as perceived.

At school, in my day, we were taught geometry to the rules as gathered together by Euclid some 300 years before Christ was born. Euclid had a very flat perception of the world around him but, much more recently, the more realistic, non-Euclidean geometry evolved which takes into account the three dimensional nature of the real world.

It is more difficult to imagine dimensions greater than three because we are accustomed to seeing all around us our three dimensional world. Whilst reading this, you are probably in a rectangular room having corners formed by two walls and the ceiling (or the floor). These features amply illustrate the three dimensional nature of our familiar world. How then can there be another dimension? Measurements taken in the plane of each of the walls relative to the datum of the floor surely give all the details of the room? Not quite so. We need the fourth dimension of time.

What has time got to do with it? Well let's suppose that the room we are in is itself moving, say as part of a train which is travelling along a railway. To fully describe the room we need to know *where* it is. To specify where it is we must give a relationship to a fixed datum. Since the train is moving, the relationship is changing. If however, we make a series of 'snapshots', we can say that at time 'T1' the position was 'P1' and that at 'T2' the position was 'P2'. This way we can determine the progressive change in relationship.

A good example of this is the movie film. A series of frames taken over a period of time show whether or not anything has moved. If it has, we could use data of the speed of running the film to calculate the rate of change of position of the moving objects relative to the stationary ones.   If the motion is a curved one, a series of dimensions would be necessary or an algebraic relationship established.

Mathematics, being the pure science that it is, does not limit itself to four dimensions - it has no limit. We will not now digress in an attempt to visualise five or fifty dimensions, although I admit to playing with the notion of multiple dimensions - just for mental amusement; this is known as the science of topology. In fact, the more one knows, the more one realises the extent of one's ignorance. However, back to the plot, what about time?

Time is a rather peculiar property. It is somewhat intangible. It is always passing, never still; like the great Mississippi, 'it just keeps rolling along'. Time is probably the greatest of nature's mysteries - not at all well understood - even by modern theorists. I can't think of a better definition of time than that given by the American scientist John Wheeler; "that which stops everything happening at once." You may say that, apart from not having enough of it for pleasurable activities, there is no real problem. An hour is sixty minutes which, in turn, is sixty seconds. What else is there to understand?

Horologists define time from the basic unit of a second which is specified as - "the tropical year for 1900 divided by 31,556,925.9747". This then leads to a problem; the definition is tied to the rate of motion of the Earth - 31½ million seconds (roughly) is the time this planet takes to orbit once around the Sun - spinning 365¼ times as it goes.

Supposing I come from another planet, 'Duffron', which has no knowledge of Earth, it is unlikely to be spinning at the same rate. What if this other planet rotates at say twice that of Earth; will everyone live at twice the pace? Quite possibly - plants do. This principle is the fortune maker of modern horticulturists; by growing flowers in an environment illuminated by artificial lights controlled by timers, the apparent day can be shortened - fooling the flowers. Their development is governed by the diurnal variations of light and so they mature more quickly. By accurate control, the moment of blooming can be precisely set appropriate to market needs.

If the inhabitants of Duffron are subjected to shorter days, will they develop more quickly? Will their society progress more rapidly than Earth's if their metabolism is geared to these double speed days? - Quite possibly. If life on Duffron commenced at the same period as that on Earth, they could have been fooling about with atom bombs and space travel while Earthlings were just discovering how to develop stone axes. Think what they may be up to now - all because Duffron spins twice as fast as Earth.

I'm thinking aloud here. I find this is a good way of probing new fields; it causes the mind to organise itself into a logical progression. By publicly stating these thoughts, others are then given an opportunity to add to the debate or challenge the fundamentals of the matter being considered.

Thinking about the fact that the horologist's definition of the second is not universal, perhaps we should consider time in the manner of other fundamental measurements as suggested in the first paragraph of this chapter. Since the speed of light (and any electromagnetic wave) is a universal constant, unit time is that required for an electromagnetic wave to travel a unit distance. Based on that notion, scientists devised an alternative formalization using the combined wavelength of a certain number of waves at a particular frequency. In the realm of rigid rules, the 'second' is defined as "The time taken by 9,192,631,770 cycles of the radiation from the hyperfine transition in Caesium when unperturbed by external fields". What a mouthful - and still the Duffrons are likely to develop twice as quickly as Earthlings.

Accepting that the speed of light *is* constant, let us reconsider the definitions of a second. The horologists based their definition on the rotation of the Earth. If the Duffrons applied the same principal they would arrive at a period half that of the Earth-second. If we consider the alternative definition based on the frequency of vibrations of Caesium i.e. 9.2 billion cycles per Earth-second (approximately), we can convert that to 'the distance travelled by light in 9.2 billion cycles' since frequency and wavelength are inversely proportional.

Since the Duffron-second is half the period of an Earth-second, light will travel only half the distance. This leads us to the conclusion that lengths are shortened and time slows down as speed increases. In this case we are talking of the angular

29

speed of planetary rotation. This is in accordance with the theory of relativity. Arithmetically, if time is reduced, distance travelled must be reduced in proportion in order to maintain the constancy of the speed of light.

If we speed up rotation more and more, ultimately time will stand still and lengths will be reduced to zero. *Time and space are not rigid – they are elastic.* If lengths are reduced to zero, space does not exist. If this is the case, travel from 'A' to 'B' at the speed of light would be instantaneous. Why then does light appear to take a measurable period to travel between two points? Is it because *we*, the observers, are not travelling at the speed of light?

If the speed of light *is* constant and nothing can travel faster than light, what about the surfer; the light-wave surfer – surely he or she can travel faster? Most people in California and in Australia are familiar with the sport of surfing. A surfer positions him/herself on the downwards travelling front of an ocean wave so that gravity draws the surfer down and, because the wave is moving, forwards. However, skilled surfers do not simply surf in the direction of movement; they angle themselves to surf along, as well as down the wave. If a 'particle' was surfing the front of a light wave but was angling along the wave, it would be travelling faster than the speed of light. Can we persuade particles to surf the electromagnetic waves of light?

The concept of time is very difficult to fully appreciate. Take, for example, the reading of a book. It happens that my wife and I often read the same book simultaneously. This is done in the bathroom so as not to waste time whilst on the toilet. I mention this incongruous fact in order to explain how it is that we can concurrently be progressing through the story although at different chapters.

If the narrative is good, one becomes absorbed in the story - 'living' the action. Here we have a situation where, for example, I can be in orbit, suffering many problems with my space vehicle, while my wife is not even on the launch pad. She's relaxing as the ground crews prepare the shuttle while I am aloft in the same vehicle, struggling for survival due to a malfunction.

A book is like a time capsule; although it can be opened at any time, progress can only be forward; one cannot select reverse - it emulates life. A baby enters into life's progression, joining those that have experienced much of life already. The newborn picks up the story of life some chapters behind its parents. Although mature people can look back at their past neither they, nor the newborn, can actually move back into time periods past. On life's journey, we all have a one-way ticket.

Because we Earthlings set our clocks according to the apparent rising of the sun, a baby born in New Zealand at 10:00 am on April 1st 1999 is half a day older than another baby born at 10:00 am on April 1st 1999 in the United Kingdom. Conversely, if two babies were born at the same *instant* in those two countries, the one born in New Zealand would be half a day younger than the U.K. born according to Earth's calendars and clocks. Very confusing! What then of someone born on another planet - Duffron, spinning twice as fast? Universal time is a must.

Speed - or more correctly, velocity - is inexorably bound with time. Just think what happens to messages sent by laser light, travelling at something like 300,000 kilometres per second (186,000 miles per second). At those speeds all sorts of strange things happen, as Einstein pointed out in his theory of relativity.

Even at more modest speeds the relationship between moving bodies has interesting effects. I used to be a jet fighter pilot and frequently flew in close formation with others - simply for fun or to demonstrate our prowess to others. We could be flying at say 400 knots within a metre of each other - the only relative motion being slight up and down movement due to my clumsiness with the controls. If the leader performed a loop or a roll, the formation would slavishly hold position on his wing tips. One would be almost unaware of the dramatic changes in attitude were it not for seeing the ground beyond the lead aircraft, at some odd angle. If, therefore, the manoeuvre was performed in cloud one would never know which way up the aircraft was.

In fact a similar technique was routinely used by the RAF to test one's instrument flying ability; known as 'unusual positions', a pilot's ability to correctly interpret flight instruments was tested by flying in a two-seat aircraft with the subject pilot shut off from the outside world by a shroud. The testing pilot would then fly the aircraft erratically to deliberately upset some of the key instruments which relied on gyroscopes. Control of the aircraft would suddenly be handed over to the subject pilot with the aircraft perhaps climbing while inverted. The subject pilot would note from the altimeter that they were climbing steeply and, if untrained, would pull back the throttles and push forward on the control column. This would be completely wrong since normality is reversed; the climb would be even steeper and, with power reduced, the aircraft would stall or maybe spin out of control - the opposite case of taking control in a spiral descent, results in a dive at increased speed.

That little digression is an example of one's brain being influenced by relativity; in that case relativity of physical attitude in relation to mother Earth. The problem lies in the conflict of signals from the eyes versus those from the ears. The eyes are interpreting the instruments leading the brain to come to certain conclusions. However, the balance system associated with the ears is sending alternative information to the brain which disagrees; only good training resolves the problem. That same problem has a parallel with sea-sickness – the eyes see the horizon at unfamiliar angles while the ears give conflicting reports regarding bodily attitude.

A more familiar example of impressions of 'relativity' (not Einstein's) is that of cars passing in opposite directions. We have all driven at say 100 kph along a road with traffic flowing the other way at similar speeds. It is quite common, even with dual carriageways, for there to be only perhaps a metre between opposing vehicles. Relative speeds of about 200 kph result. Now just imagine attempting the same manoeuvre out on an open desert. With a car coming towards you, perhaps on a curved course, at 100 kph, would you be happy to drive within a metre of it, in the opposite direction at 100 kph? I doubt it.

This readiness to travel with others at speeds of 100 kph or more on motorways can lull one into a false sense of security. If you have been driving for an hour or so along a wide, multi-lane carriageway at high cruising speeds with other vehicles around doing the same, the impression of speed is notably reduced - the mind has been conditioned. When it then becomes necessary to leave the motorway for ordinary roadways, via the slip road, we naturally slow down. It takes conscious effort, however, to slow down sufficiently. The tendency is to continue at too fast a rate. I have no doubt that this is the cause of many accidents.

Time itself also has a 'relativity' factor. "Doesn't time fly when you are having fun?" is an often repeated comment. Half an hour spent in a fun situation seems much shorter than half an hour spent say in a dentist's chair. Even long term time scales become distorted. As one ages, the years seem to pass more quickly. I believe this is because a year is a smaller proportion of your life's experience. When you were two years of age, a year was 50% of your lifetime; at age 50, a year is only 2% and reducing. The day you become a centenarian, a year will be only 1% of your life. Relatively, time passes more and more quickly.

A good analogy of Einstein's relativity is the way that sound waves are stretched or shortened by relative movement. An observer standing on a station platform while an express train rushes through will notice that the approaching train has a higher pitched sound than the same train as it recedes. This is known as the Doppler Effect, after Christian Doppler who first discovered the change of frequency caused by motion. What is actually happening is that when the train is approaching, the source of the noise is tending to catch up with the sound waves it has already emitted, thereby shortening the wavelength. If the train was able to increase speed to that at which the sound waves travel, all the waves would arrive at the same time - with a loud bang. This is exactly what happens when an aircraft arrives at the speed of sound. Conversely, when the train passes, the waves are stretched to a longer wavelength giving a lower tone.

Since light also travels in waves, this also is affected by motion; however, light travels very much faster than sound. Consequently, the effect is not noticeable at train, or even aircraft, speeds. The effect is observable with light sources travelling at a reasonable proportion of the speed of light. If the source of white light is travelling away from the observer, the waves will be stretched to a longer wavelength causing the colour to become redder. If the source is approaching, the wavelength will shorten to a blue tint.

Time also is distorted when motion approaches the speed of light. In true, high-speed relativity, clocks seem to misbehave. Our most accurate clocks are those directly based on the vibrations of the element Caesium. If we were to synchronise two such clocks then send one into space at high speed - as close as possible to that of light - when it returned, they would no longer agree. The one sent off at high speed would be slower than the stationary one. This fact is fundamental to the theory of relativity and has since been proven in actuality. If we think back to the scientific definition of time and the basis of it, we can understand why this should be so; the faster one travels the fewer vibrations of the caesium radiation will pass for a given travel distance – we have a time warp.

The strange consequence of this time/velocity oddity is that if twin brothers accompanied the clocks, the stationary twin would become older than the travelling twin. This does reflect on the possibility of space travel once engineers have mastered the problem of really high speed vehicles. Those sent out would not age as much as those remaining. This all sounds like science fiction but, as in many other situations, fact is stranger than fiction.

Thinking about the active and inactive twins leads to a little philosophical observation. The more active you are, the longer your life is likely to be - barring accidents.

In thinking about movement at speed, it is important to be aware of the difference between 'speed' and 'velocity'. Speed has magnitude whereas velocity is a vector quantity, having magnitude and direction. In other words velocity is speed in a particular direction. Acceleration is the rate of change of velocity so one can accelerate without changing speed - by changing direction.

Earlier we talked of imagining space as a cube - in fact cosmic space can be thought of as an infinite block of cubes. We have also been discussing, in earlier paragraphs of this chapter, that geometry is not restricted to straight lines in two dimensions. We have three dimensions which can be curved. With that picture in mind, we can imagine distorting these cubes by bending or stretching them, twisting them or punching a hole through them (this would involve an understanding of topology – the science of multiple dimensions). Such visions can help to rationalise what is happening when Einstein talks of curved space.

Since we have established that acceleration can be as a result of changing direction, it follows that if we travel at constant speed through curved space we will be accelerating. Gravity is the force which causes bodies to accelerate towards a centre of mass. The correlations of gravity with acceleration and space curvature with acceleration imply that gravity curves space. This leads on to the understanding of black holes where gravity has curved space to such a degree that even the speed of light is insufficient to achieve escape velocity.

To help in the visualisation of curved space let's play with our mental cubic diagram. Suppose all the lines representing the edges of our little cubes were some form of elastic metal, responsive to magnetism. If we then bring a magnet to bear, the sides of our block of cubes would be distorted. A hypothetical spherical magnet placed in the middle of the block would draw the framework towards it in a radial manner, distorting the planes of space. Since all electromagnetic waves - including light - travel along these planes, they too are curved. This is quite similar to the familiar effect of a lens placed in the path of a source of light.

It appears then that gravity acts like a magnetic lens which, to me, seems quite reasonable since all matter has an electromagnetic base; any agglomeration of matter would bring together a concentration of electromagnetic influence. This, in turn, will have an interaction with any other electromagnetic wave in the vicinity. Space is not empty; it is full of energy waves, those emanating from the source of creation – electromagnetic-gravity waves. This, we will explore more fully in a later chapter.

Since the EMG waves of space will be deflected by concentrations of gravitational energy, space itself becomes warped. The warping of space by gravity is related to the warping of time by speed since all cosmic paths are curved.

To better understand such concepts – and the meaning of time – let us reconsider nature's fundamental dimensions. It has long been established that the velocity of light (and other electromagnetic waves) is constant. Velocity has two factors – speed and direction. Speed is a relationship of distance-travelled and time. Distance and direction provide the three dimensional spatial element while time is the necessary fourth dimension. With the speed of light being constant, time becomes a variable dependant on the space traversed. The abstract nature of time can be found in the realisation that it is merely a factor of the velocity of light;

$$Velocity = Space \ (distance \ \& \ direction)/Time,$$
$$therefore,$$
$$Time = Space \ (distance \ \& \ direction)/Velocity$$

While thinking of space and time, I would like to raise the question of a possible paradox. If the Big-Bang occurred some 12 billion years ago and modern telescopes can see back some 10 billion years by looking at the most distant galaxies, the images received will be only 2 billion years after the Big-Bang. That being the case, why do they appear to be so far away? Their position, and their state of development, should be as only 2 billion years into the cosmic development.

Couple this observation with the consideration that images of close objects will be seen 'now-ish' whereas far objects will be as they were aeons ago. This will be causing a much distorted depth of focus. In fact the images appear to overtake each other. The furthermost, those that have travelled the extreme, are the youngest

35

whilst the nearest are the oldest. This takes some mental unscrambling.

Whilst cogitating that complex conundrum let's, for a moment, turn back the pages of our book once more. You may believe you have not experienced it, but you are surely aware of 'virtual-reality'; wherein by donning a special electronic helmet and spatially sensitive gloves your immediate environment undergoes a time-warp, or a dramatic relocation, or both. It turns out that this marvel of the modern electronic world has been around for centuries - long before electronics were developed. I'm referring to the simple novel, travel book, or descriptive historical account. The object of the narrative is to transpose the reader into a situation other than the fireside chair in which he is sitting. The novelist creates an environment in which his characters perform their roles. If the book is well written you, the reader, become transposed into that world of romance, terror or adventure. Your mind is temporarily unaware that you are seated with your nose in a book. You 'live' the period. You are an observer of the actions taking place wherever and whenever they take place. You are experiencing 'virtual reality' - albeit, not generally interactive.

The mind knows no bounds. This fact we will learn in later pages.

In the preceding pages, we have considered the many seemingly un-answerable questions about which the thinking person might philosophise. We have reviewed the historical relationship between science and religion, in the search for truth. A method of organising one's thoughts has been introduced followed by an exercise in thinking by looking at some of the ramifications of space and time. The object of all of this is to act as a warm-up for the presentation of a hypothesis which seeks to show that everything, whether physical or philosophical, has a common basis which leads to being able to answer all those most difficult questions and serve as a platform for further investigation.

So let us proceed.

Part 2 – The Hypothesis

Before proceeding, I wish to point out that - apart from reference to "Atom" by Lawrence M. Krauss, and checking on names and dates, what follows has **not** been researched – **deliberately** so as not to influence my own thought train. The thought process was to go right back to the physical fundamentals of the Big-Bang and from there to work through the development of matter – initially inorganic, then to organic, and so on to the evolution of plants > animals > humans. Thinking basically of what made humans human, led me to consider the development of the mind and the relationship of the soul. From this, various conclusions were drawn regarding human behaviour; how we learn, how the environment affects the development of personality, and much more; leading to my declarations relating to the All-Enveloping-Influence which some call God. **After** drafting this hypothesis, other works were then researched which have led to the realisation that many philosophers and scientists in the past have come to conclusions which are compatible with the content found within these pages although they did not appear to have discovered the link between disciplines. The boxed references made to these historic determinations are included to serve as illustrations of the substance of my deliberations.

Chapter 5

Thinking of the Big-Bang

Where do we start such a mammoth task - to find a common link between everything?

Let's start at the very beginning - The Creation.

Whereas those readers having a strong religious faith would immediately refer to the book of Genesis in order to explain The Creation, it is now well established scientifically that the creation of the universe was initiated by a phenomenal explosion - referred to as The Big-Bang. One is well aware that adherents to the many religious doctrines of this world of ours have great difficulty accepting such a revelation. An obvious comment is that for there to be such an explosion, there must have been something there to explode. This being the case, 'somebody' – God, must have created that something. Such objections are valid, given that we base our considerations on knowledge founded 2000 years ago without taking into account any subsequent updates.

According to the Oxford English Dictionary, 'Nature' is "the physical power causing all the phenomena of the material world". In other words, it is that which exists. To some religious fanatics, science is considered a 'dirty' word but science, again according to the Oxford English Dictionary, is "a branch of knowledge conducted on objective principles involving the systematized observation of and experiment with phenomena, esp. concerned with the material and functions of the physical universe". To summarize those two definitions, science looks for the true explanation of everything which nature provides. In practice, thoughts and ideas are put to objective, unbiased tests until the truth of the concept is established. That truth is then used as a basis for further exploration of natural conditions. Science seeks to establish an understanding of the whys and wherefores of nature.

An example of such development of ideas is the story of the progression of the basic theories surrounding the expansion of the universe. In the 2nd century A.D., Ptolemy declared the Earth to be the centre of the universe; no doubt due the earlier views of concentric crystal sphere's propounded by Aristotle, and to the apparent rotation of stars and planets around our globe. In the 16th century, after much careful consideration of observations, Copernicus convincingly showed that the sun was the centre of an orbiting planetary system in which Earth was found to be one of many contributing bodies. In the 17th century, Newton developed his theory of gravitation which explained how bodies could orbit other bodies. In the 20th century, Einstein expanded on the Newtonian concept to bring time into the equation with space - subsequently enabling our current, rocket propelled, exploratory adventures into the cosmos.

Each step built on earlier limited levels of knowledge - discarding that which was found to be untrue or expanding the incomplete. These steps become a matter of fact - not a matter of belief. This is how science works; building on a firm foundation.

The Big-Bang was a concept first proposed by Alexander Friedman and Abbé Georges Lemaitre in the 1920s. This was, no doubt, a result of Edwin Hubble's realisation that the universe is expanding. After all, if the process is considered in reverse such that the universe is shrunk back to where it had been, we must conclude that at some stage, the whole cosmos must have started from a point. This theory was expanded and developed in the 1940s by George Gamov with his colleagues, and has largely been substantiated by modern observations of latent radiation.

The important point here is that scientists did not invent the Big-Bang (apart from its name) they discovered a 'natural' occurrence. A fuller explanation of the physical mechanism of the Big-Bang is still being sought. This is the work of fundamental scientists - to discover what makes the universe tick, along with everything within, and to find truthful answers to the many questions which punctuate the minds of man. Later, we will return to consider the question; what caused the Big-Bang? And to indicate why, *I believe,* it did not just spontaneously occur from nothing, **nor was it created by God**.

For now, let's just accept the established fact that at the instant of the Big-Bang no material existed in the universe - only pure energy. Einstein's famous equation $E=MC^2$ (Energy is equal to mass multiplied by the square of the speed of light) shows that energy can be converted directly into mass and vice-versa. This is the basis upon which nuclear explosions release so much energy. A nuclear explosion is the uncontrolled conversion of mass to energy (although not in fact completely). Later, in the context of the declared hypothesis, I will indicate how energy and mass can ***actually*** be interchangeable.

Accepting the declaration that everything stemmed from pure energy, how has the universe developed? How has life evolved? To answer these fundamental questions we turn to the science of physics. What follows is a précis of the complex sequence of events. This is not some vision dreamt up by your author but a very brief account based on published scientific works which summarises the generally accepted chain reaction initiated by the Big-Bang.

Some readers may find themselves temporarily in an unfamiliar domain but, rest assured that this is not an attempt at blinding you with science, this is an attempt to paint a picture (or rather – a quick sketch) of how the universe actually began by the creation of matter..

A précis of *Atom* by Lawrence M. Krauss

All the material in the cosmos was not immediately created at the instantaneous birth of the universe; pure energy, in the form of electromagnetic waves, expanded rapidly from the point of the explosion. The intense heat from this massive reaction started to be dissipated as these waves grew out from the epicentre. *Even today some of these waves are still expanding, albeit they are difficult to detect. After 12 billion years (or so) they have cooled to very low temperatures.*

In the early stages, while still at intensely high temperature, perturbations occurred in the expanding waves. These random irregularities became the birthplace, first of unstable quarks, then of the positively charged protons and the neutral neutrons. Protons and neutrons were formed by collisions of pairs of quarks in the highly active state created by the extreme temperatures. Protons are stable, but free neutrons have a potential life of only a couple of minutes.

Some of the neutrons were preserved by coming into contact with protons and binding with them. Later, this duo would form the nucleus of the unstable heavy hydrogen, deuterium. Again random

---

luck came to the rescue; within minutes, some of these nuclei collided with each other to bind into the very stable helium nuclei, having two protons and two neutrons. Those that didn't bind became free protons due to the collapse of the neutrons. A single proton would later become the nucleus of the hydrogen atom.

As the energy waves cooled further, negative electrons and positive positrons were also formed. Then a remarkable turn of events occurred; in the hectic environment in which they found themselves, some of these, so called, 'particles' collided. Negative meeting positive, they annihilated each other but - thankfully - there must have been an imbalance in numbers, consequently some electrons survived. Had they not done so, the development of atoms would not have continued and the universe, as we know it, would not exist.

The surviving electrons and the newly formed nuclei continued to cavort around in the primordial soup waiting for the balance of expansion and temperature to be such that the electrons could be captured by the nuclei to form atoms. Electrons orbit the nuclei at a radius many times that of the diameter of a proton. Since the positive charge of the proton has to balance the negative charge of the electron, we find that there are as many electrons in an atom as there are protons in its nucleus. The hydrogen atom has, therefore, captured only one electron while the helium atom has two. Deuterium having only one proton, and consequently one electron, also has a neutron in its nucleus. This 'heavy' variation of hydrogen is known as an isotope.

Later in the development of the cosmos, other elements would be formed having increasing numbers of protons balanced off by co-existing neutrons surrounded normally by a number of electrons equal to the number of protons. Isotopes of these atoms (having extra or fewer neutrons than protons) have special characteristics, subsequently found to be useful to mankind in a variety of ways.

Wait a moment though, we are jumping ahead; at this stage the embryonic universe is not yet ready to leave the womb - so far only hydrogen and helium have formed. A long and complex process now begins during which, due to gravitational attraction, these atoms are drawn together into clumps. As they do, collisions begin once more which succeed in combining into more complex combinations, forming new elements.

It is worth noting the difference between the range of elements and the perhaps more familiar molecular formations which develop from them. An element is identified by the number of protons (and neutrons) in its nucleus whereas molecules are identified by the combination of atoms (complete with electrons) combining to form a new material.

The continuing collisions created new atoms such as oxygen (8 protons) then carbon (6), followed by nitrogen (7), neon (10) and others up to the heavy element iron (26), each with the same number of neutrons as protons and surrounded by a cloud of electrons. Cutting a long story short, these new amalgamations occurred during the formation of stars. Stars are the nuclear melting pots which, due to their high temperatures and pressures, create fresh material by breaking apart the atoms being drawn in by gravity then reforming them into new configurations of protons neutrons and electrons.

Nothing in the universe is permanent. Stars exist only as long as the inward force of gravity is balanced by the outward gas pressure due to the nuclear reaction within. When this balance fails, due to the 'burning' of fuel, and the gas ball collapses, a supernova occurs. This is another massive explosion akin to the original Big-Bang albeit not so all consuming. Again the combination of great heat and close proximity is the trigger for new collisions forming a further extension to the range of elements. This general sequence of events finally creates all the stable elements known to us - which is the full range possible [see appendix 'A'].

During the expansion from a supernova, conditions are right for the formation of carbon molecules - the binding of carbon with other elements to form a multitude of materials such as carbon monoxide, carbon dioxide, methanol, ethanol, and so on. The cloud of dust created by the exploding star now gravitates to eventually form a new star. The resultant further reactions enable the creation of an ever widening range of molecules. The range of carbon molecules undergo many changes, reacting with nitrogen based compounds until amino acids are formed. These are the building blocks of life from which all living plants and animals are developed.

This very brief résumé of how the universe developed serves to illustrate the remarkable fact that all materials - even living cells - stem from the energy generated at the moment of creation - the Big-Bang. Everything - dead or alive - is composed of star-dust which evolved from electromagnetic waves.

Many books have been written dealing with aspects of cosmological development, however, for those interested in learning more detail of the complex sequence, I highly recommend "Atom, An odyssey from the Big-Bang to Life on Earth ... and Beyond" by Lawrence M. Krauss. In his book, Krauss succeeds in making a dramatic story of the traumatic stages in the life of a single oxygen atom. This illustrates, most vividly, the sophisticated process of creating matter – and progressively, living cells - from pure energy.

Chapter 6

Thinking of Evolution versus Divine Design

One particular star, our Sun, was formed something like 5,000 million years ago - or more - from a 'particle' cloud resulting from cosmic explosions. There was sufficient material in the dust cloud to cause gravitational forces to draw the 'particles' together, forming a centre of mass. This concentration of mass drew in more and more material from the surrounding dust cloud. Progressively the pressure at the core built up as a nuclear reaction commenced; generating tremendous heat. Eventually the outward pressure of the hot, fusion reaction countered the inward pressure due to gravity and a stable body was formed.

The material left over formed a disc of debris around the now reacting new member of the universe. Progressively, gravity caused this material to aggregate at various radii from the parent, creating new bodies, but they had insufficient mass to start a reaction of their own. Hence the solar system was formed of a number of planets orbiting the parent star. It is worth noting that, as a result, even Earth is made of star-dust from which all the elements found here were born. As a matter of interest, the element Helium was not originally found on Earth but was first discovered on the Sun (by spectrographic analysis).

> In his book "The Ultimate Universe", David H. Levy explains that comets are balls of ice which bind together the cosmic dust left over from the many births and deaths of stars. Acting like giant vacuum cleaners these once mysterious bodies of space collect what is now called CHON particles. These are atoms of Carbon, Hydrogen, Oxygen and Nitrogen – the alphabet of life; given suitable conditions they coalesce to create the basis of living matter; amino acids. So it is a fact that we humans, being derivatives of lower forms of life, have ultimately originated from stardust floating in space.

It used to be thought that life on Earth began about 500 million years ago, as shown by fossil finds. However, it has now been realised, and proven in various locations all over the globe, that soft-bodied life was around long before, and algae type life even 1.5 to 2 billion years ago. Going back even further, it has been found that primitive life has existed in rocks for perhaps 4 billion years - maybe even before Earth was formed 4.5 billion years ago. As Lawrence M. Krauss comments in his book "Atom", credit must be given to the electron for harbouring raw energy and yet releasing it to react with materials such as amino acids to kick-start life. We will look more at how the electron harbours pure energy later.

Time then works its wonders by allowing progressive development of life - from the very primitive algae to the highly sophisticated human - along the lines discovered by Darwin and declared in his "Origin of Species". In essence, whatever the conditions favour that is the path development takes. Eventually biological plants evolved from single cells, followed by animals and from basic animals to early species of man which have then developed into homo-sapiens. Having then learned to walk upright and basically fend for oneself, mankind's next big step was the development of consciousness. We will look at consciousness in more detail later.

Currently, as a result of new archaeological discoveries, there is much controversy surrounding natural selection. It does appear evident to me, however, that as each species diversifies; weaker strains will fail thereby leaving only strong developments to survive. Man has emerged as the strongest, most progressive species on this planet – making the rest of the animal kingdom subservient to his purpose.

At this point, I wish to draw attention to views opposing the sequence of development outlined above in favour of the creation of spontaneous life.

> The Watchtower Bible and Tract Society have published a book entitled "Life - How did it get here, by evolution or by creation?" Quite understandably the content sets out to put the religious view-point but, regrettably the selected references to comments made by scientists regarding Darwin's theories fall into the

same classification as those previously mentioned regarding statistics and biblical quotations. In these references, there does appear to be as much evidence *against* as there is *for*; the general reader is being totally misled while the thinking reader tends to dismiss the comments as narrow with a distinct tendency to brain-washing.

The general thrust implies that scientists claim that life was created spontaneously ON EARTH by a miracle of coincidence. My understanding of these matters indicates nothing of the sort. As outlined above, the seed of life was not created on Earth; the temperatures and pressures necessary have not occurred here since the planet was formed. The so called 'coincidence' is not a mathematical impossibility but a mathematical certainty in the maelstrom of the primordial soup forming the birthplace of the stars. It appears that the editors of "Life-How did it get here?" – which could have done so much to further true knowledge, have followed the pre-Copernican belief that the Earth is the centre of the universe.

We Earthlings are but a minute grain in the immensity of the cosmos and must recognise that if God did create everything; He created the heavens (the whole cosmos) as well as the Earth. It is the easy way out to claim Divine Design every time an explanation of a phenomenon is not readily forthcoming.

I well recall my own father-in-law disbelieving that satellites would ever be made to orbit the Earth – 'they will fall to the ground'. "But, what about the Moon", I countered. "Ah, that is God's work," was his reply.

46

> We now have hundreds of satellites, in Earth orbit,
> sending signals for all manner of purposes; in fact, we
> could not operate modern life without them.

Fundamental scientists are trying to answer the multitude of questions that nature has presented but it is unlikely they will ever know every detail regarding everything. Little by little true knowledge grows; we must be patient. By modelling, mathematicians explore the possible outcomes of developments or ideas presented by thinking persons, whereas physicists tend to experiment to prove or disprove new theories. Archaeologists, on the other hand, deliberate on fossil finds and other historical evidence to piece together the story of life's development. Often, combinations of the various sciences are brought together in order to further extend the boundaries of our understanding. Meteorology is one sphere which relies heavily on a combination of physics and mathematics with the addition of climatological research through archaeology.

I must admit, however, that some of the conclusions drawn by archaeologists from tiny pieces of evidence seem to stretch creditability beyond reasonable limits; so much appears to be assumed from so little. The danger is that other conclusions may be drawn based on these possibly wrong judgements. As a hypothetical example, let us consider the finds of a distant, future archaeologist whose interests lie in ancient modes of transport. He discovers evidence of motor cars, railway trains, motorbikes, trucks, bicycles, and even a velocipede. It would be quite reasonable for him to conclude the following sequence of vehicle development.

Man invented the velocipede to take the weight off his legs and to move faster. An improvement in efficiency was then gained by the introduction of pedal power with the development of the bicycle. Subsequently, invention of the internal combustion engine enabled the addition of petrol driven power in lieu of pedal power – creating first the moped then the motorbike. Somebody then had the bright idea of adding a side-car which, after a little more development became a three wheeled car. Logically a fourth wheel was added to create the more stable automobile with which we are currently so familiar. This archaeologist of the future then surmised that cars were further developed to carry goods by adding a tray at the rear instead of seats – the pick-up or Ute was born. This hybrid was improved and enlarged to become the truck. This got bigger by with the addition of first one trailer then two or more. Judging from the evidence of motorways, freeways and autobahns, our researcher would conclude that, with the increase in traffic, vehicles would be travelling nose to tail along the busiest routes. The evolutionary result would be the invention of the railway where goods wagons would not be limited to one or two trailers but could be fifty or more. Our intrepid scientist has, apparently, resolved the land transport development sequence of the 18th, 19th, 20th & 21st centuries – what a big mistake!

Although this fictitious scenario has many flaws, it serves to illustrate the potential hazards of jumping to conclusions based on inadequate evidence. Even though, in this case, the sequence is wrong it is worth noting that cars, trucks, trains, and bikes **have** existed and that there was an evolutionary development, but that other factors – not here considered – caused the ordering to occur differently. This may be happening now with new archaeological discoveries being applied without full awareness of other factors. The details of the developmental trail are probably inaccurate but that does not mean that evolution did not take place – it just happened in a different order. This is not a case for trashing scientific study, merely a case for treating a new theory with due caution until proven. This flawed theory would serve as a temporary structure for further investigation which could lead to the ultimate truth of the matter.

> As the philosopher Hume noted: night follows day but day does not cause night; another influence is the cause – the spinning Earth and the proximity of the sun.

One argument put forward when considering Divine Design versus evolution is that developed life is too complex to have evolved; this view does not consider the tremendous compounding effect of multiple 'onion skins'. As a straightforward example let us consider the advances made in mechanical flight. At the beginning of the 20th century man was Earth-bound; half-way through the century, even your author was flying jet fighter-aircraft up into the stratosphere. By the end of the century we had half the world airborne in jumbo jets – in lounge suits, sipping drinks and watching films while travelling from one continent to another. The more affluent amongst us were zooming across the Atlantic at supersonic speed. In the meantime man has flown to the moon, landed, played golf, and then returned. The wood, wire, and cloth comprising the Wright Bros aircraft has, within a century, evolved into a sophisticated mammoth of the air and space. I hesitate to conjecture what the 21st century holds for the travelling public.

If a development is based on previous stages repeatedly, the end result can be far removed from the original. This is classically demonstrated by the modern digital computer; only basically able to determine '1' from '0'. Software programs layered one on top of another progressively expands that elementary ability up to the sophisticated levels with which we are now familiar. When we observe the almost intelligent behaviour of this wonder of our age it seems inconceivable that this is due to the characteristics of a grain of sand – the silicon chip; but it is.

We are currently familiar with the phenomenal growth rate of the cancer virus; imagine the massive change that would result if a tiny change took place at every stage of the development; the end result would be quite different to the initial seed. This change would have occurred very quickly; on an archaeological scale, virtually instantaneously.

It is not suggested here that this explains the apparent sudden steps in the evolution of life; merely that one must allow the mind to consider the possibilities created by repeated layers of development.

When considering whether or not life could possibly originate on Earth, it is worth investigating the primitive life-forms found deep in the ocean alongside hot vents. In these circumstances we have great heat and relatively high pressures albeit not so extreme as during a supernova. We can also cogitate on the evolutionary closeness of inanimate minerals and living beings found in a coral reef. Even the cuttlefish seems very close to being stone, with its calcified inner parts, as does the shell of a turtle or perhaps a snail. It is realised, of course, that these are examples of organic matter becoming inorganic; the opposite to the formation of life.

I don't believe that the initial elements of life originated on Earth but these examples show that there is not a total divide between organic and inorganic matter. The transformation process to life from inorganic matter is obviously very complex and is not yet fully understood; that does not mean it cannot, or did not, happen.

Those readers whose philosophy of life is based on the written word found in the Bible should not dismiss evolutionary theories as being contrary to their fundamental beliefs; simply because mankind has evolved from basic forms of life does not mean man was not created. Of course man was created – but not as a single act; man has been created progressively - and in God's image, as we shall consider later.

Chapter 7

Thinking of Waves

To discover the link between the Creation and man's activities we will need to take another look at the atom and, in particular, the electron.

We will return later, to look more closely at sub-atomic 'particles' - to show, according to my hypothesis, that 'particles' do not actually exist; that what are commonly called particles are actually waves in a special form. For the moment please accept that statement and the consequential conclusion that all matter, whether organic or inorganic, is composed entirely of electromagnetic waves. This startling revelation will be expounded upon later but, for now, we need to look at the nature of waves.

Waves are wonderful!

That sounds like a good slogan for a campaign to sell waves, but waves do not need publicity, they are all around - in many guises - to be observed, to be made use of and to marvel at.

One can sit for hours by the beach soaking up the soothing effect of watching the waves roll up to the shore, to die a death in the foaming breakers. Ocean waves ride on bigger waves, referred to as swells. These, in turn, form part of even larger waves - the tidal movement due to our moon or a Tsunami, the shock wave caused by an Earthly tremor. At the other end of the scale, the familiar ocean waves carry small waves, or ripples, on their backs.

Because the ripples, the waves, the swells, and the super swells do not necessarily run in the same direction, surface movement becomes highly complex. This complexity is increased on the grand scale by currents under the surface and weather patterns overhead. On the local scale, interference is also caused by shoals of fish, large sea creatures and by man's activities with ships and boats and the like. The beautiful, fascinating waves we've been watching at the beach are no longer simple; they are complex compositions carrying a record of much activity elsewhere.

As children, we have all been captivated by the ripples made by a stone falling into an otherwise calm pond. These mini-waves become very much more interesting when two or three stones are dropped simultaneously.

Diagram 2 – Ripple Pattern

The intersection of the ripples seems to create lines of commonality. Much technological use is made of the resulting interference effect and we shall explore such patterns created by waves other than in water.

Of course, waves are not confined to water; sound waves - although unseen - are probably equally familiar. These enter our ears to be converted to nerve signals for onward transmission to the brain, bringing to us the sounds of nature; warnings of danger; conversation with those around us, and the harmonies of music.

Sound is the result of vibrations transmitted as fluctuations of air pressure. These vibrations can vary in magnitude (loudness) and in their wavelength (sometimes referred to by the inverse relationship of frequency). Since the speed of sound - in standard conditions - is constant, a variation in wavelength results in an inverse variation of frequency; the longer the wavelength the lower the frequency of vibration. Dangerous seismic waves are very similar to sound waves since they are pressure waves generated within the Earth's crust. We are now all aware of the grumbling and groaning of Earth's tectonic plates which result in disastrous Earthquakes transmitted to buildings by these potentially powerful waves.

In the meantime let's wallow in the wonderful waves of music. Music is basically the manipulation of a sequence of sound waves. Laboured as that statement may appear, in the brains of great composers - be they classical or contemporary - such sequences are arranged to bring much pleasure to the world; pleasure through harmony.

What actually is harmony? Harmony is simply a sequence of waves which fit together.

Plucking, bowing or hammering of strings and skins create vibrations. The reeds of saxophones or woodwind instruments and the lips of players of brass instruments all make vibrations. The pipes of organs vibrate in proportion to their length. These vibrations are tuned, and often amplified by the particular instrument, to generate characteristic sound waves. The vocal chords of a singer vibrate at different frequencies according to the muscular tone applied.

A perfect tone results from an unblemished wave which rises smoothly to a maximum, descends back to the 'at-rest' level, then continues to descend smoothly to a minimum - the precise opposite of the maximum - before returning to the 'at-rest' level. Technically, this is known as a Sine Wave [Diagram3].

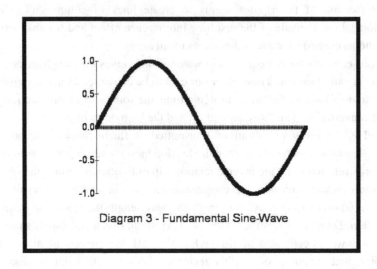

Diagram 3 - Fundamental Sine-Wave

We could produce a similar wave but at a proportionally higher frequency (shorter wavelength) - giving a higher pitched tone – as Diagram 4, below.

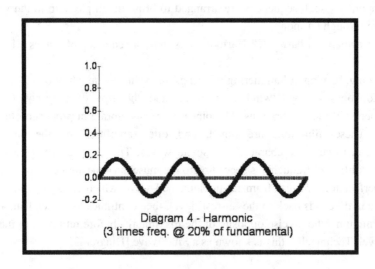

Diagram 4 - Harmonic
(3 times freq. @ 20% of fundamental)

If we now add the two waves together - superimposing the higher frequency wave on the fundamental - the sine wave becomes modified to give a new, harmonious tone - as shown below [Diagram 5]. The higher frequency wave, being a multiple of the fundamental is referred to as a harmonic - in this case the 3rd harmonic, because its wavelength is a third of the fundamental.

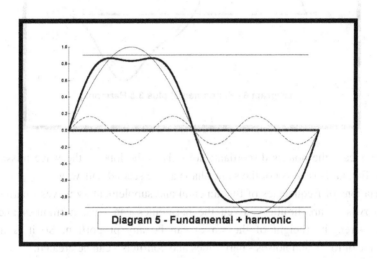

Diagram 5 - Fundamental + harmonic

It is harmonious because the new wave shape (that shown bold) is evenly balanced; with the positive lobe (that above the line) and the negative lobe (that below the line) identical. The result, however, has a different tone because the shape, although smooth, has a prominent kink creating a double peak. We could go on adding other waves at various compatible harmonic frequencies. The end result would be a complex wave which, as a sound, would produce a pleasant tone.

On the other hand, if the higher frequency wave had been introduced out-of-phase (not passing zero at the same instant as the fundamental) or it had not been at a compatible frequency - say 3½ times instead of 3 times - then the result would be a deformed wave, as Diagram 6 below.

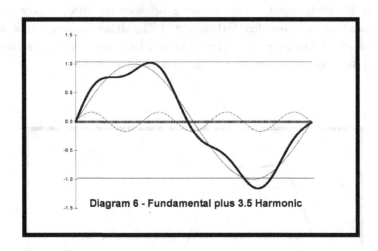

Diagram 6 - Fundamental plus 3.5 Harmonic

In this case the note is discordant; not only is the lobe of the wave misshapen but the first half-wave is not the same shape as the second half-wave.

Variations of frequencies of fundamental and supplementary waves - harmonics or otherwise - are limitless. Even the harmonics can have harmonics and the relative strengths (height of the wave) can be any proportion. So it is that a multitude of tones and noises - harmonies and conflicts - can be created.

An orchestra, comprising many different kinds of instrument, has to be very meticulous in ensuring they are all tuned to the same basic frequency otherwise discord will result. Usually, the leader will play a particular note, perhaps an 'A' which has a frequency of vibration of 440 Hz (or cycles per second), and all other instruments adjust their tone while playing the 'A' note until a synchronised fundamental is achieved. Although the various instruments each have a different timbre, they will be in harmony.

When the orchestra plays, the listener will hear the combined sound of all the waves emanating from the whole range of instruments present. Because they are basically in harmony and the notes chosen by the composer are also compatible - due to being 3rds, 5ths and 7ths (third harmonic, fifth harmonic and seventh harmonic) - a pleasing sound is heard. If someone should play a wrong note - not in harmony - the waveform of the combined sound will be distorted and the sound will be unpleasant. I should admit, at this point, that the harmonics of the diatonic scale are rather more complex than the above diagrams suggest since the fundamental is not a single wavelength. However, the principle remains. We will see that this concept of harmony or discord of sound waves does not only apply to music but, much more significantly, to other kinds of waves which affect general human behaviour. It is also interesting to note that savants, those people with special natural abilities, display their unique characteristics in drawings, mathematics and music; the *drawings* above were created by *mathematical formulae* to represent *musical tones* - waves being the link.

The work of the great classical composer Johann Sebastian Bach is considered the foundation of all modern music; in the early 18th century he created the cornerstone from which all western music evolved. It appears that this came about due to his mathematical approach to the art of composition. He even devised the chromatic scale where the octave is divided into 12 semitones having equal progression. Although this is musically not precisely correct, Bach recognised that this mathematical approach enabled the different tones of the variety of instruments required to form an ensemble to be combined and give the harmonious tone required. Music and maths are shown to be harmonious.

Roughly fifty years after Bach led the way Beethoven lifted the development of music by another significant notch. In "The Stream of Music", Richard Anthony Leonard explains how, unlike his predecessors, Haydn, Mozart and others, Beethoven didn't generally build his major works from individual elements, he sculptured passages to fit his overall concept – he thought in pictures, pictures of the finished work. This revelation is quite similar to the manner in which a good engineer conceives the solution to design problems; his visualised final solution is made realistic by the detail development of the component parts. Here, we discover a link between the mental concepts of art and technology - spatial cognition, a mathematical concept.

Most readers, although familiar with the visual delights of ocean waves and the harmonious pleasures of musical sounds, are less familiar with the most common form of wave, the electromagnetic wave. Specific knowledge of these unfelt, largely unseen attributes of universal existence is usually restricted to those with a technical background but, when one thinks about them, we are all aware of their existence in many forms. We have radio-waves, micro-waves, X-rays and, of course those which are most well known - light waves. These are all examples of electromagnetic waves; the difference between them being their frequency of vibration.

The part of the spectrum of electromagnetic waves with which we are most familiar is the visual range - that which we call light. Ordinary white light is a spread of frequencies giving a blend of colour, from Red to Violet, which, if split up, give us the beautiful rainbow. Laser light, on the other hand, is all at one frequency, or colour. The range of colours of light can be compared with the range of sounds we call music; they can be used in subtle combinations to please our visual senses similar to the manner in which harmonious sounds bring music to our ears.

The many other non-visible waves make up the electromagnetic spectrum. Infrared waves, micro-waves (used in cooking, communications and radar) and radio waves have lower frequencies than light, while Ultraviolet waves (which cause sunburn), X-rays, and the very dangerous Gamma Rays, have - more energetic - higher frequencies than light. Fortunately, most of the harmful rays which abound in space are filtered out by our atmosphere leaving mainly the lower frequency radio waves. So it is that, just because we cannot see or feel them it does not mean we are not being influenced by waves.

For instance let us consider brainwaves. We usually consider this term in reference to an idea born within one's cranium; however the brain does actually generate electromagnetic waves. At this stage we should consider the brain's similarity with a modern device - the radio. Because the human brain relies on the transmission of electrical signals through the synapses between neurons, it acts to some degree like a radio; the electrical activity emits a weak signal in the form of an electromagnetic wave.

A radio *receiver* (or a TV) will *transmit* a weak signal on the frequency to which it is tuned. (This is how detector vans can check whether or not viewers in the UK are using an unlicensed TV receiver or how a listening bug can be detected in a room.) Similarly brain transmissions can be received by sensitive detectors placed on the skull and sometimes by other brains in close proximity. The brains of pigeons and certain other life-forms are tuned to the Earth's magnetism to assist in navigation. This has been proven by applying magnets to specific parts of the skull to create upsets. This ability is not confined to animals; some humans also have this attribute. Personally I found, as a fighter pilot, I could always find my way home - even without sight of the ground - I was, when younger, a human pigeon. This magnetic sensitivity also accounts for directional confusion if one moves from the northern to the southern hemisphere (further confused by the unfamiliar position of the sun). Your author has also had this experience having spent most of his life in the northern hemisphere before migrating to Australia.

Since we do not only receive signals but also transmit them, albeit normally over short distances, this creates a personal aura; faintly displaying the activity in one's brain. So, it appears there is more to the brain than the obvious; not only does it control our basic functions and our thinking, it is interacting with other brains around it. Let us then look at the development of the brain, and then begin to consider the mind and its link to the surrounding environment.

Chapter 8

Thinking of the Brain

Thanks to Watson and Crick, the world is now very much aware of the importance of genes; the string of DNA which control the development of all living bodies whether plant or animal. It is the change of the variable combination of genes over millions of years which has given the mechanism by which Darwin's evolutionary principles do appear to work.

Plants have genes which enable development of the physical structure from the most simple - and oldest, single cell to the highly complex, often beautiful floral displays found in many gardens; and the majestic trees found in tropical rain forests of Brazil or South-East Asia; as well as the great coniferous forests of North America. Plants are, however, limited by the lack of mobility but - more importantly, the lack of ability to think; they do not have a brain. The genes do, however, control the method and timing of reproduction; they dominate the procreation process.

Animals have genes which, besides developing the physical structure to include mobility, introduce brain controlled behaviour of the bodily operating systems. This enables the introduction of muscle controlled features such as the heart, lungs and the motor operated arms, legs and bodies etc. Most importantly, the brain acts as the centre for coordination of the senses. This centre of observation develops into the centre of control; taking full advantage of animal mobility. When it comes to reproduction – the maintenance of the particular species, the animal brain uses its sensual information to be selective; by choosing well developed, healthy partners, healthy offspring become the likely outcome.

As the variety of animals evolved, the brains of some species developed a second layer of activity. Besides the inherent gene programme enabling essential bodily functions including reproduction, the brain gained an ability to remember events. Memory allows the development of learning skills which, in turn, leads to the ability of the brain to interact with itself - it creates a mind. Animals have not, however, developed a mind of sufficient ability to fully appreciate self because they lack the influence of language.

Humans have additional genes which enable the development, not only of the bodily structure and brain controlled operating systems, but also of a very sophisticated mind. The human mind developed the means to communicate within itself - the language of the brain; it 'talks' to itself - enabling thought process to take place. This language of the brain appears to be basically a picture language; a picture built up of interlocking electromagnetic waves. We are all aware of the cliché *a picture is worth a thousand words;* this way the brain can deal with a vast array of information. These abilities have evolved as 'man' has progressed from the ape-like form of prehistoric times to the present.

Let us investigate the relationship between brain and mind.

The general familiarity we all now have with computers, particularly the PC, leads us to make a comparison with the brain. Although the two do not function in quite the same manner, there is a similarity. The similarity is the ability of both to receive and process a mass of information from a variety of sources and then to utilise the result to instruct other 'devices' what action is required and how it should be performed.

Of course a computer does not have a mind of its own; it needs to be programmed so that it can respond in an appropriate way. The brain is similar in this respect; it needs to be programmed; initially it is the genes which set the programme for basic functions – including growth and ultimately sexual reproduction.

The programming of computer and of brain has multiple stages. At its most basic a digital PC knows only 'on' or 'off''; which can be interpreted as 'Yes' or 'No', 'Go' or 'Stop' or any other choice of one from two. The computer's basic Operating System – which can be likened to the human gene programme, has been programmed to develop this very simple task into relatively complex procedures by thousands of repeated actions carried out at great speed. The next, and subsequent stages of programming progresses those procedures to eventually provide a format - such as Windows - which the operator finds convenient in order to select other programmes of actions. You, the user, may then operate these programmes in whatever personal manner suits you - you personalize it.

The human brain can still be compared with the PC. The embryonic brain is initially programmed to carry out basic procedures such as controlling blood flow, the muscles, and ultimately (after birth) breathing. This programming stems from the genes. The second stage of programming comes from the mother (or more specifically, the person who actually carries the foetus – perhaps a surrogate mother); she starts the process of developing a mind by induction. After birth, as the brain progressively gains sufficient operating experience and memory to 'talk' to itself and *think*, the embryonic mind is formed. We will go into how that happens shortly. The personalization of the brain occurs with the development of the mind. Due to experience, the mind itself becomes 'programmed', but this too is a multi-stage process which we will discuss at length.

An aspect of the brain - which is different to that of a computer – is the physical manner in which it works. When brain cells are excited to respond, they do so by discharging an electrical pulse across a gap (a synapse); this is very similar to the old fashioned spark transmitter which was used in the early days of radio. It appears that the resulting weak transmission caused by the inter-cell signal can be received by adjacent cells – this is induction.

Induction is a very important process; without electromagnetic induction there would be no dynamos, no alternators, no transformers - no electrical system in this modern world of ours. Basically, induction is the phenomenon where a fluctuating electromagnetic field will create a sympathetic charge in any close conducting medium. However, it is much more subtle than that. A variety of electromagnetic waves can be induced simultaneously such that they all become added together to create a new profile. Even more subtly, if the waves being received are fundamentally similar but out of phase with each other, a travelling wave will be created which can move at a greater rate than the originals. Since, in a vacuum, all electromagnetic waves travel at the speed of light, this means that the travelling wave can move - contrary to the current laws of physics, at a speed greater than light-speed.

Another aspect to be considered is the relationship of speed and time. At Earthly speeds there is no problem, but as one approaches the speed of light the rather difficult to perceive phenomena of the slowing of time becomes apparent. The nearer to light-speed one travels, the slower the clock until, at light-speed, time stands still. This being the case, all EMG waves are in unison – there is no time difference from any wave to any wave, instantaneous actions can occur. An analogy could be that of two trains travelling in the same direction, at the same time, alongside each other on adjacent tracks. This is an occurrence frequently observed by commuters approaching or leaving busy city terminals. We now have parallel worlds, travelling at speed where, theoretically, a passenger could transfer from one train to the other. Train 'A' would lose a 'particle' while train 'B' would gain a particle, instantaneously. In the cosmos, since EMG waves occupy all so-called-space, a 'particle' or signal, can move virtually instantaneously from anywhere to anywhere.

The cumulative effect of all of this is that one section of the brain can be influenced by those around it - at great speed, virtually instantaneously. I suspect that the potential of high speed travelling waves can account for dramatic moments of inspiration - even somewhat disconnected from current thoughts. This broadcast activity may cause adjacent cells to respond until an aura is created around the active part of the brain, resulting in a new activity such as an invention, a musical composition; perhaps a great work of art

Man has created two basic types of computer, the digital and the analogue. The former handles discrete pulses, somewhat like the pulses of Morse-code used in telegraph systems of the past, whereas the latter processes continuously variable values such as rising and falling temperatures. It could be that the brain cell with its attendant synapse acts like an analogue to digital converter. The chemistry of the cell is being influenced by analogue signals from sensors such as the eyes, or feed-back from say the heart together with other induced signals from adjacent cells. When the activity reaches a certain level, a signal is sent to adjacent cells for further processing; that signal is a digital response.

The accumulation of emanations from the various regions of the brain creates a general aura of waves around each of us. The characteristic of these waves is complex and variable. As we saw earlier, waves can vary in frequency and in strength but, most importantly, in shape. The basic shape of a pure sinusoidal wave becomes modified by the addition of a multitude of superimposed waves, in and out of phase, received from within and from without the brain. This mix started even before we were born.

> For some years, engineers have been developing robots capable of carrying out ever more complex tasks. The ultimate aim is to create a machine that can

> think for itself and make on-the-spot decisions.
> Modern robotic space-probes have this ability to a
> limited degree. The limitation in emulating the human
> brain is primarily due to the enormous capacity
> required to truly learn and remember.
>
> Mark Tylden, a radical robotics engineer, seems to
> have realized a significant feature of the human brain –
> the interaction of various sectors. Using an analogue
> rather than a digital system he links various elements
> into what he calls a 'nervous net'. This enables much
> greater interplay with minimal complexity.

There is a further influence which, quite subtly, has a bearing on the mind. If we think back to earlier chapters, it will be remembered that the original electromagnetic radiations from the moment of creation are still present, albeit rather weak. These waves - modified by the secondary radiations created by cosmic developments such as supernovae - have an 'All-Enveloping-Influence' on everything and everybody. Inanimate objects simply receive these waves and re-radiate them at a modified frequency. Animals, with their brains predominantly programmed by their genes, are also unable to interpret them in any significant way although we are aware of what has been termed 'animal instinct' by the more advanced species. Numerous incidents have been recorded where animals have sensed danger and responded as a result. These happenings probably stem from the superior sensitivity of animals to scent, pressure and other physical factors, but they illustrate the ability of their brains to assess the implication of the combined input and to initiate action accordingly.

The developed brains of the human, however, are potentially influenced by any background radiations whether stemming from localised transmissions or from the cosmos. With effort, one can tune-in to the unadulterated characteristic of the original waves from The Creation (the Big-Bang) to become aware of their underlying attributes. Most significantly, whole sections of the human brain receive, and respond to, transmissions created by other sections within the brain.

So how does the mind learn to think - how is it programmed? The start point is the conception of a child. During the development of the foetus, the brain initially evolves according to the programme dictated by the parental genes. Later in the pregnancy, the aura of the mother - created by her brain transmissions and carried by her blood - begins to impart a characteristic to the electromagnetic activities of the child's brain. Even though the 'All-Enveloping-Influence' has been present from conception - providing a smooth basic waveform on which to build the new individual character - the constant proximity of the motherly vibes will probably overcome this weak external signal.

The result is that, whatever characteristic the mother has, this will be progressively applied to the developing mind of the unborn. If the mother's waveform is in harmony with that of the waves emanating from The Creation the resulting characteristic of the developing mind will also be in harmony. If, however, the mother is in conflict with the 'All-Enveloping Influence' then the offspring will have a mind which is also in conflict.

Up to the moment of birth, the father will have little influence other than by the initial gene programme imparted at the moment of conception. Once the child is born, if the father is close and gives loving attention to his recent offspring - particularly by touch - then, progressively, his characteristic will have some bearing. It is worth noting that, from the moment of birth, the child is not quite so constantly close to the mother - thereby reducing, slightly, her influence. Given that the parents are themselves in a harmonious relationship, the child's character will progressively assume this common aura. Conversely, if the parents are not really compatible then the child's characteristic will have conflicting components - which I shall call 'conflictics' (as opposed to *harmonics*) - added in, changing the waveform to an uneven cycle; causing some confusion and uncertainty.

It would appear that none of these imposed characteristics, with added harmonics and/or, 'conflictics' are permanent; their affect slowly dies away. It is constantly necessary to reinforce the desired features in order to give a more lasting durability. It also is evident that the first features assimilated take precedence over those subsequently applied. If one thinks about it; having learned something one either builds on it or, if subsequently found to be wrong or unsuitable, it has to be un-learned as well as re-learned. From this we realise the importance of coming into the world in a suitable environment and learning important attributes early in life.

Let us recap. The functioning of the brain relies on electrical pulses being transmitted between neurons as a result of signals being received from the various sensors in the eyes, ears, nose, mouth, and by touch and by induction from adjacent brain cells. The basic functions have been set by the parental genes, as has the development of the rest of the body. Although areas of the brain have become associated with particular functions, the effects are not limited to those regions. The activity of one part of the brain will, due to electromagnetic induction, cause associated effects elsewhere. In the human brain, this activity leads to 'thinking' which, in turn, back-feeds into the mind. From this an awareness of self develops.

Self appears to be the established characteristic waveform predominant in the mind of a person; this we will call the 'personality code'. If other waves, introduced through the senses, have forms that fit to this general characteristic then they will be drawn in, building strength to that aspect. Conversely, if it is found that the introduced wave does not fit, the resulting conflict will cause rejection. It is this acceptance or rejection which is the awareness of self. This only begins after a few years of experience of life, the time-scale being dependent on the stability of the upbringing.

> The philosophers John Locke (1632-1704) and David Hume (1711-76) shared the opinion that it is only from experience that our knowledge of anything outside ourselves can be ultimately derived, whether it be our own or somebody else's. In answering the question of whom or what "I" am? Hume concluded that "I" am a bundle of sensations.
> *The story of philosophy* by Bryan Magee

It should be stressed, at this point, that the frequent references to characteristic are not indicating some arbitrary nature of a personality; it is the actual shape of the waveforms governing the various sections of the mind which have been developed through experience and training - it is the sum total of the many harmonic wave-forms that have been received, countered by any 'conflictics'. Training is, of course, organised - often repetitive, experience which strengthens particular features of the characteristic waveform. Experience, having been presented in the form of harmonics, which are added to the primary waveform, or 'conflictics', which are rejected, cause the maturing characteristic to be aware; aware initially of self and progressively of right and wrong; good and bad; the favourable or undesirable.

As mentioned earlier, the original electromagnetic waves from the Big-Bang - The Creation, are still in evidence. These are rather weak due to the passage of time. Other waves emanating from a multitude of cosmic events such as supernovae and re-radiated from inactive bodies such as planets, comets, asteroids, and even all the surrounding Earthly matter also approach us from a variety of directions. These cause complex waveforms at the points where they come to be added together.

Inductance having caused the aura of the mother (and to a lesser degree, the father) to be combined, together with the later addition of experience, creates a unique characteristic – the personality code. This characteristic will be allied to a specific combination of ethereal waves which become the stable anchor - the soul. The soul is, therefore, the juxtaposition of ethereal waves coming together from every quarter, which combine to form a characteristic which is the same as the basic personal characteristic which has been built up so far. The soul is the programmer directing the mind while the mind is the software of the brain.

The degree that the soul is harmonious to the original waves emanating from the Big-Bang will be a measure of the person's spirituality. These original waves, being the 'All-Enveloping-Influence', are a direct link to The Creation.

After the child's birth, the wave characteristics of a caring father will have an increasing tendency to modify this newly developed characteristic, finding a new combination of ethereal waves. The event of the premature demise of the mother would, of course, allow the father's harmonics to become entirely dominant. On the other hand, an absent father will not be in a position to offer any contribution. An orphan will receive the sympathetic frequencies of the foster parents. If these are generated by genuine love and affection, they will be tuned in and cause favourable adjustment to the characteristics of the child's mind.

In my book "View of Life", I related the account of my wife and me fostering a newly born Nigerian baby - conceived of a totally different ethnic background to our typically white English upbringing. Initially, he didn't even know he was black; after all, the only other humans he knew were pale skinned - apart from a brief

> *visit by his father which frightened him greatly. It is very significant that he was about seven or eight months old before his thinking ability was able to recognize the difference between him and us. It was a moment that will live in my memory for ever. Whilst I was holding him, he saw himself in a mirror. I could almost see his brain working as he repeatedly looked first at the mirror image and then at myself, as he realized that the person next to me; the pale one in the mirror, was himself, the black one. Then, brilliantly, he thought of a way to confirm that his colour was different to mine by lifting his hand to see that it too was black. His father's subsequent visits were never quite so frightful.*

This event was a classic example of how a mind develops with the input of experience gained from the senses. It is also interesting to note that the discovery did not trouble him but, when he next had a brief visit from his father, he was less troubled by what he saw. What a bundle of joy he was until, due to our own problems preventing us from continuing, he was placed with a couple having no sympathy, no harmonious vibes. His previously cheerful disposition became sullen and uncooperative; a clear example of how alternative frequencies having been superimposed on the fundamental can either enhance or distort the original characteristic.

From this premise one can understand how it is that siblings developing within a harmonious environment will emerge as stable characters. Whereas those from casual associations will be un-tuned and unstable until some new influence focuses the soul to a particular frequency having a strong fundamental characteristic. A point being made here is that the soul is a 'floating' point; one that is basically anchored but can be moved by changes in character. If, due to life's experience one's character adjusts, then it will key in to a new combination of ethereal waves. This change will draw the person towards others of similar characteristic or further away from those in conflict.

As we progress towards adulthood, we have associations with many new acquaintances, notably one's teachers, or one's religious leader. The aura surrounding these contacts will be different, by varying degrees, from one's parents and the parental influence will potentially be diluted. If, however, the new input is fundamentally on the same wavelength, it will add to the character already established and so we progress towards developing a sound personality. If the parents have no scholastic interest, the child will experience a conflict of characteristics. It then depends on how allied the teachers' characteristics are as to whether or not the child will absorb the formal teachings. The same applies with the priest, or imam, or other religious leader.

Perhaps some casual connections will be on a different wavelength altogether. If our fundamental character is well tuned to a particular frequency, or wavelength, little distortion will take place because we will be strong enough to reject or dismiss the conflicting element. On the other hand, if our upbringing has been an unsatisfactory hotchpotch of associations, then no firm characteristic will have been established. The unfavourable transmissions will then be able to deform our fundamental, causing a change in our personality. It is for this reason that prisons breed criminals since the aura within is not one of harmony with the 'All-Enveloping-Influence' but of harmony with a distorted fundamental.

On the other hand, a couple finding themselves attracted to each other by more than sexual desires are being drawn together by the interchange of harmonics progressively modifying their characteristics - they are beginning to think alike by enjoying each other's company. Their love for each other gains momentum as the shape of their keys become closer to fitting the same lock.

In his book "The Consolations of Philosophy", Alain de Botton portrays 'A contemporary Love Story' in which a man and a woman – having quite different interests, meet for the first time on a long distance train journey. They are drawn to each other – even before they speak, but why? The author points out that early in the 19th century the Polish philosopher, Arthur Schopenhauer gave a name to this moment of 'man's' irrationality; he called it 'will-to-life'. His reasoning was that all animals – humans included, have an inbuilt need to ensure the continuance of their species; as a result, something within responds to the possibility of a suitable mate.

Schopenhauer did not, of course, have any knowledge of gene theory but it was basic gene programming that he was alluding to. The will-of-life is the innermost layer of the proverbial onion skins of the human computer we call our brain. The subsequent layers of characteristic applied, first by parents then subsequently by acquired experience of life, mask this underlying fundamental programme which influences our whole outlook on life.

It is important to realise that the basic need to promote our species as triggered by our genes may or may not be totally harmonious to our developed character. Some of us show very strong sexual tendencies which incline to overpower our more rational traits, while others subdue these basic urges in favour of more, shall we say, 'worldly-wise' behaviour. The point to recognise is that we may not be consciously aware that our attraction to someone of the opposite sex is based on the gene programme designed to ensure the survival of the species.

While the production of offspring is biologically possible, a bond exists between united couples whether or not they are temperamentally suited. Only if the couple in question exchange harmonious vibrations, and allow their characteristics to adjust to suit total compatibility, will a satisfactory relationship endure. All too often we meet couples who bicker and snap at each other yet are 'good-in-bed'; their relationship will not survive the difficulties ahead. Some other couples do manage a reasonably stable common life so long as they can bond with their children and their grandchildren but their own inter-relationship is often only lukewarm.

It is rare for a couple to truly bond – to become as one. The implication is that not only are the basic gene programmes in harmony but the developing characteristics are truly sympathetic such that they grow in unison. One aspect of this association is that it should be triggered before the two individuals have developed a firm personal character; otherwise they would have to 'unlearn' to engage in common vibrations. Late marriages are likely to suffer difficulties due to the developed characters being too inflexible; they are unable to adjust their personalities to properly harmonise. Both parties would also benefit from the experience of family love in their own parental homes so that the concept of harmony is inherent in their upbringing.

For the coupling to succeed beyond the child-bearing age, without the help of the emotional ties of descendants, they must share their thoughts openly; conversation is an essential ingredient. Talking of life's problems and pleasures continually adjusts the characteristics to be in sympathy. This does not mean that the man becomes as the woman or vice versa nor must it require total subservence of one to the other; it does, however, generate understanding. Consideration of others is a vital element in any harmonious relationship - especially transient ones.

If two people are fortunate enough to achieve this harmony, they will share the simple joys of life; the delights of autumn colours, the excitement of flying off to a new venue, looking across a candlelit table at the object of one's affections, and simply relaxing in front of a fire on a cold and windy night. More importantly, a truly harmonious relationship enables the sharing of problems; discussing unfortunate, maybe even disastrous, situations – without apportioning blame, will help to alleviate the pain and concern for the future. Sharing the bad as well as the good strengthens the bond and prepares well for the future – ensuring the continuance of the harmonious union.

Conflict between such partners reduces virtually to nil since their characteristics mirror each other and they become almost one. How can one argue with oneself? Interactive thinking is employed to deal with problems as they arise - the remaining difference between them enabling complimentary input.

Your author is fortunate to be party to a union having full harmony; a living example of true bonding, such that in over 55 years of marriage no argument has ensued, but we've had plenty of drama to test our relationship while putting much spice into our life together.

In *The God Delusion* Richard Dawkins draws attention to the current interest in 'Cultural Variants' – or 'Memes', by various researchers but notably Susan Blackmore who, in *The Meme Machine,* pushes Meme theory further than anyone. Memes appear to be what I have termed 'developed sub-characteristics' – wave-forms with beneficial attributes which, like building blocks, combine to create stable sectors within the mind.

The act of thinking focuses the mind on to particular wavelengths, selecting or rejecting depending on one's strength of view (or character). This applies to couples, partners in business and as individuals. This way a thinking person will develop a stronger personality than a person who simply accepts whatever is fashionable.

Restating these observations we see that, if we all think and act in a similar manner, a smooth, trouble-free life will result - but wouldn't it be dull? To add colour, we create additional elements and superimpose our ways onto the accepted norm. If these traits are in keeping with the general characteristic of those around you, you will be in harmony. The addition of your input can add tone to the lives of others. If, however, you choose to ignore accepted good principles - having little consideration for others, your additional element will be in conflict, causing an unfavourable disturbance.

The conflictics of modern life increasingly prevent the development of suitable fundamental characteristics to attract long term harmonious relationships. Nature's lowering of the age of pubic development also appears to create desires before the youth has gained sufficient worldly experience for his/her embryonic characteristic to develop a stable base. Relationships are engaged in for purely sexual reasons, whether or not the person realises the depth of nature's procreation programme. The environment of early divorce and homosexual relations creates confusion when the young adult searches for what is normal. It all comes down to waves; those that harmonise with the All-Enveloping-Influence or those that harmonise with life's conflictics.

The conclusion drawn from this picture is that your existence need not be dull and lifeless so long as any contribution one makes is in proper relation to the swell of life generated by humanity as a whole. Remember, that little ripples can happily travel on the back of waves which are themselves riding the big swell.

So it would appear that waves do not only have a bearing on harmony in music but truly on sexual relations and life as a whole.

Chapter 9
Thinking of the Common Denominator

Let us now go back to the beginning, not the beginning of the book, the beginning of everything - The Creation. The moment when the Big-Bang created the electromagnetic equivalent of the mother-and-father of all tsunamis; a great outburst of energy creating a super swell expanding in all directions. This was the moment of conception of the whole universe; not yet the birth but the conception.

For a brief period, the radiation of this superheated energy was consistent in all directions, but then random perturbations caused waves within the waves. Progressively, an agglomeration of waves formed, creating 'islands' of increased intensity which drew to themselves primordial gases - the birth pangs of the cosmos. These embryonic developments became the birth of the first stars. This activity itself caused the re-radiation of waves as celestial hot-spots developed. Now, some 12 billion years after creation, these hot-spots have generated into great galaxies of stars which, in the scheme of things, give birth and suffer deaths (supernovae) creating ever more waves amongst the waves.

The original outward surge still appears as electromagnetic radiation but the resulting waves are made less distinct due to cooling and the confusion caused by the emanations from the multitude of secondary and tertiary sources of the galaxies. These 'new' centres of radiation are located all over the cosmos causing waves at any point to be received from every direction.   Imagine, if you can, the incredible complexity caused by waves radiating from billions of points in space - in every conceivable direction. At any point in space-time there is a unique combination of interacting waves.

This is the crux of the hypothesis. At each and every point there is a wave-form element direct from The Creation with harmonics or 'conflictics' superimposed upon it. They combine to form a characteristic peculiar to that four-dimensional location. We humans, with our well developed minds, have also developed a unique characteristic waveform. As with a key fitting a lock, each human mindful wave-form fits a celestial ethereal waveform; our mental radiations are tuned to a particular combination of cosmic radiations. The point in space-time where this combination occurs is what we choose to call our soul.

Our personal characteristic or soul is that which directs the mind. In turn, the mind causes the interaction of the various elements of the brain. The analogy of the PC would suggest that the human brain is the *processor*, the mind is the *software program*, and the soul is the *programmer* making the basic decisions as to how the 'system' should work.

The initial influence of parents on a developing child establishes the basic characteristic which, if firmly implanted, will create a strong fundamental wave as a link between mind and soul. With a strong waveform, the mind will lock-in to the appropriate ethereal waves and thereby establish the soul in its cosmic space-time location. From here on, as the child matures to gain independence from his/her upbringing, the soul will relocate according to empirical tendencies. The degree by which the characteristic is likely to change will not only depend largely on the firmness of the foundation learning but also on the severity of experience in life.

One must remember that, besides the guidance emanating from the soul, be it strong or weak, the mind is subject to many pressures from other minds. As discussed earlier, harmonious influences will add colour to the existing character. On the other hand, influences in conflict would distort the form of the character wave unless rejected by superior strength.

I think we have all met people who dominate a gathering; they have a strong characteristic. On the other hand, many of us are either shy or reluctant to force an issue; lacking that characteristic strength. That does not mean that the character is actually weak; perhaps that personality code contains a larger element of consideration for others. Those people who *do* have a weak characteristic are likely to follow, blindly, those of strong character.

I must apologise for continually referring the 'characteristic' – this is fundamental to the whole concept of the universal link. May I remind you that it is not some hypothetical notion; it is an actual combination of electromagnetic waves which individualises the person - or even the object, giving off vibes.

Let's reconsider the analogy of ocean waves. There are great swells, upon which waves develop – the waves are added to the swells to give a particular character to the whole. We must remember that the waves and the swells come from different sources. The swell may be from the south-east while the waves perhaps are coming from the north-east. Onto that combination of nature's ocean waves, man now adds further complication by the passage of a hypothetical ship, creating new waves generated at the bow and at the stern. The end result can be quite complex due to the multiplicity of sources. The difficulty in appreciating the reality of characteristics is the fact that we cannot grab them and preserve them in a bottle – they are intangible. Nevertheless they are real.

When thinking 'pure' thoughts - those thoughts not influenced by worldly matters or by the multitude of conflicting waves from all quarters of the cosmos, we are tuning our minds to the original electromagnetic waves from The Creation.

It seems evident to me that God is in fact the pure energy manifest in these ethereal waves. As explained earlier, these waves have created all matter; 'the heavens' and the Earth, the oceans, the clouds in the sky, the land, the fauna, the flora and mankind - including even our DNA.

Man *is* made in His image - from His waves.

We now can see how it is that God is in all of us; we are made of the same 'stuff' as He – electromagnetic waves.

Although this statement may shock many who have deep religious faith, it should not be interpreted as an irreligious declaration. Quite the contrary, given that all religions should be based on truth rather than myth, this hypothesis gives a firm foundation for all to commune with God by whatever doctrine is preferred. To deny physical realities is to engage in mysticism which contributes little towards the pursuance of truth. It is, therefore, constructive to have a foundation on which to build. With this structure in place, God is no longer a vague spirit, usually visualised as a bearded man, but appears as pure energy accessible to all.

How can we access this source of spiritual strength?

If we are alone, in some remote place, there is no other brain emanation close enough for us to receive. One can then think clearly; the mind can be focussed on whatever one desires. To go out on a clear night, whilst free of city life by being on a remote hill or plain, to view the spectacle of pin-points of light in the sky, is bound to inspire. One thinks of the origin of those lights only to realise that each one is a galaxy of stars or even a cluster of galaxies. Their great distance from Earth then becomes apparent and one can only gaze in awe.

For me, this triggers an interesting chain of thought. The vast numbers and vast distances involved in the make-up of the universe lead to me cogitating on the wonders of mathematics, just as one might be enraptured by some great literary masterpiece or renowned work of music. It is good to have a feel for numbers, just as we have a feel for the words appearing in written text.

Let's, for a moment, consider the comparison of maths with literature. Maths, or more properly - mathematics, gives the majority of people the shudders. Memories of hours of boring multiplication tables or visions of unintelligible algebra are what spring to the minds of those who haven't seen the light. "Why should I learn all that junk?", "Who wants to know that 'Y' equals so many 'Xs'?" These are the questions that sons and daughters ask their mums and dads.

I could equally ask, "What difference does it make whether I am 'paid' or 'payed'? Or "Should I wear a pair of 'socks' or a pair of 'sox'? The point is that, without learning the basics, it is very difficult to discover what the subject is really about. To properly understand language (which, I hasten to add, I do not), it is necessary to read good literature and poetry and so develop an appreciation of the subject. Experimenting with words can help a writer acquire good sentence construction, which gives satisfaction to the writer and leads to the greater pleasure and understanding by the reader.

We all know the advantage of language - it is our most important means of communication. Touch, visual expression, and scent are alternatives but these are usually limited to the emotional moments of love or drama. Our language of speech and writing is used to convey our ideas, our instructions and our needs. It is also used within oneself to review our hopes, our fears and our desires. Without language, we could not pass on knowledge, in fact, we could not engage in abstract thought - we would be no more advanced than the Neanderthal Man.

Mathematics is simply another form of language; a form more suited to shapes and patterns. Learning arithmetic is akin to learning to spell correctly - it is a necessary early step towards much greater delights. Having mastered that first phase, moving on to algebra, logarithms, trigonometry, calculus and other branches of this science can be compared to learning good phrases, linking them together with correct punctuation to form meaningful sentences and then creating masterful compositions. Progressively, by reading and writing, one becomes familiar with techniques of word-craft so that not only can the words themselves be better understood, but also the reader is able to enter the mind of the writer and really 'feel' where the author is taking him.

Similarly, with practice, the mechanics of mathematical techniques become second nature allowing the mind to assimilate the underlying problem and so probe for a solution. It is not everybody that wishes to explore the logic of a problem, nor does everybody wish to become another Shakespeare. It is because most of us do read and write that we - to some extent, practice the art and develop some understanding of those that make it their career.

The same cannot be generally said for mathematics. Owing to the educational system, which fails to show the point of arithmetical exercises, most young students see the subject as one which requires a pass mark in the examinations rather than becoming aware of its future potential.

Regrettably, I am not an accomplished mathematician but I did develop sufficient interest to open my mind to what lies beyond. "So", you might ask, "what *does* lie beyond?" For me, my basic training opened the door in my mind to reveal all manner of wonders; spaciousness, movement, force, patterns of logic, even colour and curves; size, shape, style and so much more - all this because I think in pictures rather than in words. Isn't this wealth of vision generous reward for the effort of gaining a foothold in the 'dreaded' maths? I think so.

Music is another universal language; the language of harmony. Music conveys feelings; excitement; anger; love; and hate. Again, it is most desirable that the fundamentals should be learned early in life so that a proper understanding and a true feel can be gained for its wonders. My characteristic has been endowed with a harmonic sympathetic to mathematics and its associated logic for which I am grateful since it is so fundamental to understanding 'what makes the world tick'. Unfortunately, a lack of appropriate early training has left me with a weakness in the literary and musical spheres. This is where many students in the past (and perhaps currently) have missed out.

In my opinion, the target of educationalists should be to generate the enthusiasm necessary for the student to pursue these basic subjects and so develop the appropriate harmonics in his/her characteristic wave, rather than mechanically learning to pass examinations.

I seem to have digressed but that little excursion does serve to further illustrate how one's characteristic is built up from a series of harmonics relevant to different aspects of life - so, back to the plot. If we are not alone, but in a crowd, other people's thoughts are creating faint waves all around us. When the cluster of people around you is of random purpose, this collection of weak signals just serves to cloud one's mind. This is why one cannot think clearly in a crowded place. On the other hand, if those around you are there for common purpose, their thoughts are likely to be in phase with each other - including your own, and summate to a strong signal influencing the whole assembly - creating an aura.

> The renowned psychiatrist Carl Jung developed a controversial theory based on what he called *the collective unconscious*. It is probable that he was referring to this aura; that weak electromagnetic radiation we all emit, as if we are mini radio broadcasting stations, combining with others of like mind to create a more forceful general aura.

I offer some everyday examples of collective aura. In the first case we have sports supporters. Let's imagine two soccer teams locked in a frantic challenge for 'The Cup'. Supporters of team "A" obviously all agree that they wish to see their team win - they are all on the same wavelength. Each individual spectator receives sympathetic mental transmissions from others around him. This heightens his own emotions; his brain effectively says "I must be on the right track because everyone else agrees".

The combination of thousands of people all tuned in to the same objective generates an emotional wave of such power that it affects the bodily function of the players. When one of the team gains possession of the ball and progresses forward towards the opponent's goal, the increased wave of support gives his adrenaline an added surge: he runs faster, ducks and dodges more sharply - increasing his chances of scoring.

Of course, the supporters of the opposing team are generating alternative waves. These give strength to the defenders but, since they are naturally on a different wavelength, they have little direct negative effect on the attacker unless the wave of emotion is actually directed against the attacker personally, perhaps due to bitterness. The balance of numbers of team supporters is obviously relevant; this is probably why the home team usually has an advantage - simply because it is easier to travel to a local match there will be more spectators with an affiliation to the home team, creating a greater combined aura of support than that of those travelling from afar.

If you were a supporter of the winning team, and were in amongst many other fellow supporters, you would feel the satisfying warmth of success. This stems from the magnifying effect of the aura of those around you. Your characteristic has been temporarily modified by the harmonic associated with the recent pleasurable experience of the match. If, however, you found yourself amongst a group of supporters of the losing team, without engaging in any communication, their aura of disappointment would be in conflict with your so recently modified characteristic. This will moderate the effect - almost as if you are thinking "That could have been our team that lost". These sensations of harmony or conflict have been due entirely to the transmissions emanating from all the brains around you.

The second example of collective aura is that of attending a church service. In the regular service, the objective of the minister is to create an aura which is harmonious to that of God. By praying together, all minds are being focussed on the subject of the prayer. The emanations from those around you will be in sympathy with yours, building the strength of the appropriate harmonic. The singing of hymns further aids the development of the tuning of all minds to a common wavelength - that of The Creator.

Ministers who are well practised in the technique of evangelical preaching are able to develop this common tuning to a degree where the minds of the members of the congregation are whipped up into fervour. In engineering terms, this is known as resonance, where a signal is bounced back and forth having an accelerating effect on the subject(s) involved. Consider a pendulum swinging steadily backwards and forwards due to the input of a tiny pulse of energy from a spring or battery. If that synchronised pulse was to be increased just slightly above that necessary to overcome friction, the pendulum would swing further and further - out of all proportion to the input. If this is continued to excess, breakdown will occur.

A good example of the effect of resonance, with which most readers will be familiar, is the screech of a loud-speaker when a microphone is held in its area of broadcast. Sound enters the microphone, is amplified and then transmitted by the speaker back into the microphone and so on, around and around, instantly raising the volume to distortion levels. This can be damaging to the equipment as well as to the ears.

The human brain can work in a similar manner, sending a signal to a sympathetic colleague who responds to your support by returning (quite involuntarily) a strengthened signal to you which causes you to react similarly. With many like minds around you, the magnifying effect is much more pronounced. It only requires leadership to co-ordinate the minds to achieve this effect. This is where the evangelical preacher focuses his effort. If the end result is a greater awareness of one's ability to tune in to God's ethereal waves, the outcome will be satisfactory.   There are, however, people in this world of ours that use this same technique to further their satanic aims. This is how so called religious sects develop to the stage where unseemly acts are performed. In either case, there is a danger that persons of less than strong character may be unable to cope with the experience of resonance. The receipt of a strong harmonic - being further boosted by sympathetic signals - can overcome a weak characteristic. It may distort it to a degree beyond which it cannot recover. We now have a warped mind.

Other examples of collective aura can be the cause of difficulties in human society. For instance, riotous mobs develop as a result of a critical mass of people - who are on the same wavelength, being incensed by actions of others. If this group of people are firmly locked into a common frequency, a subversive signal can readily influence mass behaviour. Whereas this is deliberately used to disrupt political meetings and the like, it can accidentally cause trouble at, or near, sporting events. The malicious intent of a hooligan team supporter, or an anarchistic person at a political rally, can be received by others in physical proximity. If they are tuned closely to the thought wave of discontent, they too become infected. This 'cancer' has the power to grow if crowds of like-minded people are within the sphere of influence.

Waves of terror are an obvious extension of the above phenomenon. In this case, the perpetrators are buoyed up by sadistic vibrations from their leaders who have probably - quite deliberately, infected their minds by progressively tuning them to their own venomous wavelength.

An appreciation of the probability of the re-tuning of one mind by those around can lead to an understanding of the context in which The Holocaust occurred. A few persons, such as Hitler and Himmler, together with their henchmen had, what might be termed, unsociably acceptable notions - including the elimination of the Jewish race. They progressively introduced a harmonic in those around them which was sympathetic to their cause. The Hitler youth, for example, were brainwashed right from their early associations with the regime. These subjects, after a suitable 'incubation' period, would not be too averse to instructions to carry out minor unsociable tasks - perhaps arranging for all Jews in a region to be identified.

After becoming used to the actions in hand, their characteristics stabilize in the appropriately distorted manner. A further input to that harmonic would strengthen it in relation to the person's original characteristic and allow him/her to increase the severity of his actions - perhaps to gather together all Jews in his/her area. As they carry out these tasks, they would infect their subordinates. Of course fellow officers, likewise engaged, have developed similar harmonics, setting the mental pendulum swinging a little stronger which, in turn, influences more members of the regime. This process continues until totally outrageous, sociopathic behaviour has become the norm. It is then quite difficult for colleagues and associates to resist this evil mental wave.

Many officers and men of the 3rd Reich, who had originally developed quite normal personalities, were progressively affected in this way by the apparent 'normality' of frequent outrageous behaviour. In more recent times, a similar situation developed at the American prison at Guantanamo, on the island of Cuba. Here degradation and torture became the norm.

Environment plays a big part in creating the aura of normality in what can develop into a criminal regime; for example, the wearing of a common uniform – particularly one which incorporates jack-boots, leather belts, and other dominant paraphernalia, together with close association in barracks; all would build a sub-characteristic of superiority. In these cases officers of the regime could readily, at least temporarily, be re-tuned to become quite evil. Peer pressure is very strong in the mind of an under-developed character. Those re-tuned members of the regime with families would, no doubt, have a dual personality – a separate sub-characteristic which related normally to their loved ones when they were in their own home environment, and the brutal, sadistic sub-characteristic which became manifest whilst carrying out their antisocial activities..

When the war ended, these same individuals, if previously brought up in harmonious surroundings would, when isolated from the noxious waves of their colleagues, and changing their uniforms for civilian garb, would probably revert to normal behaviour, progressively expunging the conflicting waves from their temporarily deformed characteristic. Those that didn't had been permanently infused with a blind belief in their leadership – still convinced that the policy was just. These latter characters would be truly of a warped mind.

> *As an aside, just think; if DNA profiling had been available during the Hitler era, every German citizen would probably have been subjected to a blood test in the search for those of Jewish descent. If so, it would probably have been found that even Adolph's great idol, Wagner, had a small percentage of this supposedly dreaded fluid.*

As another example of collective aura, we should examine the prison system. The regime of prison officers breeds a culture of brutality, again due to the aura of their associates - many of whom are on the wrong side of the fence. This 'aura' is aggravated when the establishment is in a remote location - away from the stabilising effect of a normal social environment. It is no wonder that criminals breed criminals in our prisons when, not only those incarcerated are of a rebellious frame of mind, but those charged with their supervision often find difficulty in resisting the pressure of the inherent harmonics of conflict – conflictics.

> *It is my opinion that strenuous efforts should be applied to establish alternative punishments for all but the most serious wrongdoers. Correction of the faulty characteristics should surely be the objective rather than brutal punishment which appears to create the breeding ground of antisocial behaviour.*

Crime waves are born from similar roots. A disadvantaged group of society, such as the inhabitants of the ghettos of some cities, re-educates its members to eliminate inhibiting "normal" thought waves. By substituting waves of justification via the 'needs-must' syndrome, they can readily be persuaded to carry out felonious acts; after all, survival is a basic requirement of all living species. If those around you normally indulge in mischievous activities, their mental waves will be frequently received and absorbed to the point where you too have been tuned to similar thought waves. How often do we learn that someone with a weak personality has been drawn in to the activities of a gang or group? In areas suffering low employment and the associated deprivation, social rules are readily modified. Only those of strong character will be able to reject the pressure to succumb to unfavourable harmonics.

> *Perhaps, for that reason, potential employers looking for*
> *strong characters should pay attention to those applicants*
> *from less fortunate areas.*

A less critical group of people who are affected by a collective aura are the followers of fashion. It is the 'in' thing to wear this or that because to not do so resists the aura of your peers. Because they are of like mind, adherents get sucked in to the general undercurrent due to constantly receiving 'vibes' from those around. These vibes can cloud one's own ability to think. This is not restricted to fashion per se but to other trends in art or literature. For instance, it has been known that art lovers have been drawn to admire an abstract painting which was subsequently found to be upside down, or to fawn over a pile of bricks on the floor having no artistic form whatsoever. These, so called lovers of art had not been thinking clearly, merely 'going-with-the-flow'. It is only a small step before the minds of such 'followers' are influenced by their confederates to the point that they become 're-educated' and begin to see the world from a distorted angle - all due to waves received from close by.

Art – whether it is sculpture, painting, or music, is an expression of feeling; of emotions; it exudes the aura of the creative artist's moment of inspiration. What then is the difference between a painting and a work of art? - Vibrations. An artist can create a painting that may be pleasant to look at but an inspired artist transfers his emotions via his brush to the paint on the canvas – the work becomes imbued with his thoughts and passions. This aura of passion is then re-radiated to anyone able to attune to the vibes.

83

Leaders in the world of espionage have taken advantage of group aura; successfully recruiting spies from intellectual groups. University graduates, who, believing in their own perceived superior intellect, have been known to become politically active and, due to bouncing the 'pendulum' back and forth with those of like mind, become candidates with suitable harmonics to have their characteristics re-tuned to allow them to carry out treasonous acts. Cambridge University in England became the breeding ground for an infamous group of spies whose warped minds betrayed western policies to the eastern block.

All these cases clearly suggest the potentially insidious nature of collective aura and the resulting danger of resonance. There is no doubt that our minds are transmitting signals and that these signals, if repeated can progressively modify the general characteristic of a person. The harmonic is 'being learned', becoming a semi-permanent feature of the principal mind wave and, therefore, re-tuning the soul.

To tune to God's waves one must, therefore, adopt one of two methods; engage in collective assembly in a recognized church or temple, alternatively seek solitude as far as possible away from distracting waves so that clear thinking can take place. It is a matter of choice probably based on one's upbringing.

A much more pleasant effect of resonance is that of a well performed musical concert. I must admit to being a fan of Andre Rieu who has the ability to temporarily re-tune the minds of members of his audience. For those readers who have not enjoyed the pleasure of his concerts on television, DVD or, better still live, I should explain that Andre is the charismatic leader of a first-rate classical orchestra backed by singers. He leads from the front with his violin, launching into a very diverse repertoire, interjected with comic verbal introductions. The audience is soon captivated and can be seen to respond to the variable moods of the music being offered. One has witnessed dramatic pieces literally bringing tears of emotion to grown men, while the light-hearted selections bring joyous excitement to all present. The resonant response of the audience to the musicians raises the aura, affecting the performance to lift the concert to even greater heights; it is quite magic. Members of the audience become intoxicated – without imbibing, to the point where otherwise disciplined ladies and gentlemen begin dancing in the aisles and generally displaying uncontrolled behaviour.

The above example of resonance generating waves of emotion truly underlines the importance of harmony – in music *and* in emotions; this is never truer than the cases of genuine love between partners.

Chapter 10

Thinking of Conscience

In the above paragraphs, we have seen how readily a mind can be re-tuned from the previously developed characteristic to an alternative. If the re-tuning is due to, say, established educational courses, the outcome will probably be beneficial to the recipient. If, however, the 'adjustment' is due to unfavourable associations, the subject person can take a very wayward path. How does a person decide which side of the tracks the current experience lies?

You may well conclude that it is simply a matter of education; knowing what might cause harm to one's fellow man; knowing the law. But how *does* the brain *know*? It all depends on the environment of one's development.

A classic example of the very strong influence of the environment of upbringing is explored in the final chapter of Carl Zimmer's book "Evolution". The associated television documentary dramatises the scenario to great effect. In the state of Illinois, Wheaton College is a Christian school with a strict Creationist policy (all professors have to sign a document stating that they accept the literal Bible account of man's foundations being the creation, by God, of Adam and Eve). The students studying biology are, nevertheless, required to be introduced to the alternative evolutionary doctrine propounded by Charles Darwin.

These conflicting ideologies of creationism and evolutionism cause much disagreement with parents of students since *they* are entrenched in the belief of the declarations found in the book of Genesis. Those indoctrinated into the literal word of the Bible cannot concede an alternative to God having directly created

Adam and Eve. Adherents' characteristics have been formed totally sympathetic to that concept - any other explanation creates a 'conflictic' which is rejected. Scientifically trained students have progressively had their characteristics tuned to accept the principle of evolution; consequently they are prepared to consider any evidence in its support. The two parties to the argument view the case from different sides of a fence because their characteristics are tuned to two different environments.

In these modern times, we are very conscious of the threat of terrorism and the willingness of some devotees of the Muslim faith to carry out dastardly deeds. The underlying attitude - the ingrained belief, of such adherents is born from the environment of their upbringing. If, from the moment of birth – and even slightly before, the parental vibes declare the virtues of following Allah and glorifying the heavenly domain awaiting martyrs, the developed characteristic will accept such behaviour as being *right*. It is, therefore, reasonable that such a person would have no reason to doubt the correctness – even the desirability, of carrying out a lethal terrorist act **- *in the name of Allah.*** This is fundamental to *his* belief. In his own environment he is not a criminal, he is someone to be admired. The accumulated waves have developed a characteristic which is in harmony with the religious teachings he has followed.

In his book *The God Delusion*, Richard Dawkins quotes Voltaire as follows: 'Those that can make you believe absurdities can make you commit atrocities.'

[This does not imply Allah is absurd, but refers to the absurdity of heavenly martyrdom.]

Bertrand Russell similarly comments 'Many people would sooner die than think. In fact they do.

> The whole first section of this book, *The Link,* draws
> attention to the importance of being in the habit of
> thinking – of mentally challenging that which has been
> offered; all of us should practice that habit.

Initially, a foetus knows nothing but, while still in the womb; basic brain waves are being imbued with the mother's aura. After birth this tenuous character is bombarded by information during the immediate post natal period. The flimsy, transient character waveform is being constantly upgraded by the addition of harmonics received from those closest. If these are basically in sympathy with each other, the character will develop a firm basic form. One must remember all elements of the mind's characteristic are not entirely permanent; they decay unless re-stimulated by appropriate mental activity. Any particular wavelength being reinforced by learning, or by association, will gain strength, causing favourable adjustment to the overall characteristic.

So it is that we all develop our early character as a direct result of the environment of our upbringing. If then a new association introduces a wave at an unfamiliar frequency, it may or may not fit the pattern already established. If it *is* compatible, it will add colour and depth but if it is not on a suitable wavelength, conflict occurs. It then depends on the strength of one's character whether or not to reject it or adjust one's character to accept it.

What has not been established is whether this new input is considered 'right' or 'wrong'. If the circumstance of one's upbringing was stable and law abiding, the judgement would normally be socially acceptable in the generally approved sense, since the characteristic has the appropriate profile. However, we must bear in mind that the 'socially acceptable environment' can be one of extreme religious bias. The developing personality accepts the teachings of The Pope, or the Jewish faith, or Islam, as being the norm – that which is right. Alternatively, the character in question could have been developed in a stable but criminal environment. In this case the judgement would be appropriate to the maintenance of the accepted lifestyle, as dictated by this profile. A person born to the gang-life portrayed in West Side Story or the mafia-riddled Chicago of the 20s and early 30s has different standards of conscience to those of the ordinary man in the street.

The 'norm' changes with familiarity. For example, to be a jet-fighter pilot is generally considered to be the height of manhood, earning the admiration of those in less taxing occupations. However, as an ex RAF night fighter pilot, I am well aware, those around do not hold you in awe simply because, they too are jet-fighter pilots or are closely associated with flying: this is their normal environment. Similarly, famous people are not generally overawed by other famous people because this is their norm.

To visit an exotic country introduces the mind to many new 'vibes' which fade as one returns to the familiar neighbourhood. If, on the other hand, the visit was more than transient, those new vibes would become absorbed into the characteristic behaviour - they would become normalised. So it is with those who are unfortunate enough to live in an ethnic city ghetto, their norm is likely to be quite different to those living in Palm Beach or a luxury penthouse on Millionaire's Row. What is abnormal to one person is quite normal to another. It all depends on how the shape of one's waveform compares to another's.

As a person, it would appear that one is only able to recognise what is normal or what is not, by what fits, or what conflicts, with one's existing characteristic. This then is my explanation of conscience - the test as to whether or not an introduced thought (harmonic or conflictic) will fit to the waveform of one's existing character. Whether that appears morally right or wrong depends on the circumstances of the earlier formation of the person's characteristic.

It is probable that a strong character develops due to continual sympathetic stimulation - by the neurons becoming almost 'hard-wired', as a result of consistent activity, or at least by the development of preferential synapses. Since the activity of the mind of a well developed character will tend to follow similar courses, the aura will be synchronised to a particular juxtaposition of ethereal waves. Resonance will then take place with the pendulum bouncing electromagnetic signals back and forth - between mind and soul. The character has now been firmly established.

Close relatives share a bond born out of their root development. Having stemmed from a local branch of the same family tree, their minds have been imbued with similar fundamental waves - their souls are tuned to very similar characteristics - they are said to be 'close' even if geographically far apart. Although they may have subsequently deviated from the path of their forebears, experience having added different harmonics to the fundamental, the foundations laid cause them to rely on the same basic thought waves. It is for this reason that a sound upbringing is so important. As with any building or bridge, a firm foundation is a prerequisite to a successful outcome.

Twins brought up together will tend to retain the same characteristic since their experience during development will be similar, any changes will be due to the same harmonics. On the other hand, siblings who suffer a marked difference in early experience will naturally receive quite different harmonics causing a divergence in their characters.

True lovers (not simply sex partners) are drawn to each other because they can tune into each other's mind and share similar thoughts. By the very nature of their sympathetic thoughts engaging into the same wavelengths, resonant harmony ensues. The sexual act becomes more meaningful since this occurs only when both partners are inspired - rather than incited - to indulge. This inspiration again comes from thinking harmoniously. In a long lasting marriage, or close association of friends, the individual characteristics filter out undesirable elements and absorb the desirable harmonics until they approach the same profile: this is being 'at one' with each other.

Solitary contemplation is free of the influential thought transmissions of others; giving the opportunity for clear thought. Uncluttered by irrelevant waves, the mind can focus upon the waves of choice. The direction a free mind will take relies on the fundamental waves, modified by the recently experienced waves. If that recent experience is in phase with one's ever developing characteristics, harmony will result. If, while in that geographically isolated situation - while free of Earthly vibes - one tunes in to the fundamental ethereal wavelength, one will be in contact with the Creation. True communion with God will have resulted since the original, creative energy waves – which are what society calls 'God' (whether it is admitted or not), will be evident.

Ancient man was generally unaware of the world outside his own immediate environment. His mind was uncluttered by technical developments, politics, religious doctrines, or any other matter apart from his needs for food, shelter and the inbuilt need to further his species. As such his thoughts were very basic; his wavelength was very close to the cosmic fundamental. Consequently, his conclusions were very sound; he was able to project along avenues which had fewer deflections. The sages of the past were the thinkers who tuned-in to primary waves, possibly to the extent that, in some cases, they were able to genuinely see into the future - the seers.

Old remedies, along with old wives tales, are no doubt developments of these ancient deliberations. How often have modern scientists found that their latest medical wonder is merely a variant of something that has been used for years? The wisdom of ancient times, drawn from access to uncluttered waves, is still percolating down through the ages but is shrouded by a mist of modern diversions. We should not lightly dismiss that which has been passed down from generation to generation.

Savants, drawing on the waves produced in the innermost mind, are able to put depth into drawings; without special training, some can mentally calculate very rapidly; others are able to play music accurately after hearing a piece only once. Perhaps the innermost region of the mind, normally clouded by our currently sophisticated thought patterns, is clear of such diversions – clear to develop fundamentals.

Teachers (or parents) achieve much better results if they can bring their students into their own minds; by persuading them to tune into their frequency - to think with them. This surely is why personal tuition is much more effective than book learning. My own educational experience highlights the effect of being out-of-phase or in-phase with the teacher/lecturer. At my war-time secondary school, where we were treated as mindless kids who had to have knowledge forced into us, I was not interested in learning anything. Then I moved to a technical college where the students were treated as persons in their own right; my interest in education was awakened. This was the college's method of attempting to synchronise the characteristic waves of themselves with their students. It is not every lecturer (or every parent) that has the ability to convey his innermost thought waves. It is also not every student that is able to be receptive to the transmissions of others. This really is what is meant by being a good listener; it is more than simply hearing what is being said, it is flexing the mind to allow the reception of other waveforms.

Astrology is frequently the subject of much ridicule but it could have a basis of truth - albeit long lost. If one visualises this Earth, sparsely populated by ancient man whose mind is free to receive emanations from the cosmos, where are these waves coming from? As previously established, the cosmic fundamental waves originate from The Creation but all bodies, the stars and the planets, re-radiate these at modified frequencies. The closest bodies to Earth are the Moon, followed by Mars, Venus, and the other planets. It follows that the proximity of sources of waves at the time of birth - when the mind is most susceptible - may well have an influence on the initial waveform. As previously mentioned, that which is absorbed (learned) early in life has a greater tendency to last, simply because it is the basis on which new experience is added. It would be expected, therefore, that the disposition of 'heavenly' bodies at the moment of birth *could* have an influence. It is, however, quite another matter to believe that anyone could *know* in what manner a new-born has been affected by the position of the moon or the planets.

None of the above examples of thought patterns are conclusive in themselves but, collectively, they appear to me that, although diverse, they exhibit a common thread. Try to visualise powerful waves emanating from The Creation - the Big-Bang - being perturbed by random chance, creating hot-spots which, in turn, generate new waves all intermixing throughout the developing cosmos. Over aeons the hot-spots become galaxies of stars, some with planets. Some of these planets receive waves beneficial to the creation of life. Life itself is, therefore, based on an extremely complex admix of waves. Waves re-radiate so that this life-form develops in an amazing variety of forms. Still the waves spread out from - and into - every mind, whether human or animal.

The sheer complexity of these waves hides the basic structure. It doesn't seem possible that initially simple electromagnetic waves can develop firstly into material (this we will look at more fully shortly), and then that some of this material can become alive. It is even more remarkable that living beings are guided in their development by waves but yet, the more advanced species i.e. man, is able to decide for him/herself which path to take; all this through the medium of electromagnetic waves.

Suppose I was to state that it is now possible to make dirt take you to the office, to school or to the theatre. Would you readily believe it without some significant thought? Well let us see how this might occur.

Many years ago, man discovered that all dirt and rock was not the same. For instance a red ore was found which could be melted to form a liquid that could be shaped as it cooled into a hard metal - iron was discovered. Similarly copper, lead, aluminium, silver, gold and others were progressively added to the inventory. After many decades, it was found that iron could be combined with carbon and other elements to make steel. Steel could then be used to make sheet which could be folded and bent and pressed into a sort of box shape. This box was then supported on steel shafts called axles to which were fastened steel wheels.

The sap from a particular species of tree was found suitable for processing into a material we call rubber. This could be shaped into a hollow ring which was fitted to the rim of the wheels. By blowing air into the hollow tube, the rubber ring was made flexible while being able to support the weight it was carrying. These hollow rings we call pneumatic tyres. Originally these boxes on wheels were towed by horses but then it was found that the metals which man had learned to manipulate, could be fashioned into a complicated hollow lump with all manner of bits and pieces inside. These pistons, crankshafts, valves and the like were able to slide, rotate, open and close until by squirting some special vapour inside and causing it to explode; a steel shaft was forced to rotate. This drive shaft was then connected to the wheels of the box to make it move - without a horse. This we call an internal combustion engine.

Skins of animals were processed and then fashioned to cover seats placed in the boxes for people to sit upon for a ride. Sand, soda ash, and limestone were fused together to make a hard transparent material we call glass which we then use to keep out the cold and wet while looking out. Much more complicated processes were developed to make plastic materials out of the dirty black liquid which issues from certain holes in the ground. These plastics are shaped into all manner of features to finally make this box into an automobile. Copper and aluminium were processed into wires and sand was converted into chips to feed computers so that the whole thing could be controlled. We use the automobile to take us to the office, to school, and to the theatre.

For all that the above description is painfully dragged out; it is a gross simplification of how dirt and rock can be converted into a sophisticated modern motor car. This exercise, however, serves to illustrate how something basically simple can, through many repeated processes, become a sophisticated, energy filled, active 'device' with some ability to think for itself. So it is with cosmic waves progressing, via a very complex process, to material, and from that inorganic material, via a very improbable process, to living bodies. Some of these living bodies then progressed to thinking persons. A few of those thinking persons, such as Copernicus, Newton, Einstein, Watson and Crick, and others worked out how this has all come about – well almost, we just need to link it all together; to connect the sophisticated, computer-controlled, automobile – or whatever, with the dirt from which it originated.

Of course, dirt itself is comprised simply of waves.

The end result of repeated processes is so far removed from the origin that it is very difficult to believe that the one stemmed from the other but – you must have faith, faith in mankind's ability to resolve complex problems. The solution to one problem frequently opens the door to solve another problem. This line of development is expanding exponentially leading to unbelievable levels of understanding.

We must not underestimate man's abilities; equated to all others in the animal kingdom, man's intellect is like the universe compared to a grain of sand. I recently witnessed an outdoor, night orchestral concert which set off a philosophical train of thoughts. Just looking at the whole scene of an audience focussing on a stage of musicians illuminated by a complex array of lamps and laser effects, I thought – no animal does anything like this; they don't make entertaining music, they don't gather together to watch others, they don't have any ability to create the desired setting. What does a cow know of lasers? - Nothing. What does a horse know about amplifiers, microphones and loudspeakers? - Nothing. Would a wombat know how to design and build a temporary grandstand? - Certainly not. Can an electric eel design complex electrical circuits? You must be joking. Chimpanzees are clever but can they make a trombone or a violin? - No way José. All these animals have existed far longer than modern man yet have made little progress.

Of course, all of these tasks are well beyond the abilities of animals, but in the comparative short time that man has existed, he has developed into a thinking-human with a vast range of abilities. Before WW2 most homes did not have a telephone – why bother, there was nobody to ring; because they didn't have a telephone; it was the chicken-and-the-egg syndrome. Today we can telephone anywhere in the world while moving in a car, flying in an aeroplane, or cruising the ocean. Our pre-war scratchy recordings lasted about a minute, were fragile and weighed far too much. Now memory sticks can carry many pieces of music in a size similar to that of a man's thumb. Before the war, passenger aircraft were small, with very limited range and woefully slow - now, the current populous is able to jet around the world, leaping from continent to continent while enjoying a film, accompanied by refreshment to suit one's palette. What about email; my parents could never have imagined that I could write a message in Australia to be received virtually instantly by my brother in the U.K. Man-made satellites feed information down to Earth to appear on televisions or be received by data-processing computers. None of this was available when I was a lad – not even the television. Initially wireless reception was by way of the crystal set which required earphones due to not being powerful enough for a loudspeaker. Only just before the war did we progress to the radio, which, by the way, was operating by courtesy of the thermionic valve – not the transistor, nor the chip. Look where we are now; it is quite incredible what has been achieved within this person's lifetime.

Although man's academic development has been a natural – albeit a remarkable, progression, we owe a tremendous debt to the great thinkers of the past who lifted our perceptions above the dull acceptance of the middle ages.

These great leaps in man's intellectual abilities are largely to the credit of the great minds and their inspired thoughts. Probably the greatest was Einstein. To illustrate his motivation, I reprint a couple of paragraphs from Marcelo Gleiser's book *The Dancing Universe.*

[Einstein]
"The most beautiful experience we can have is the mysterious. It is the fundamental emotion which stands at the cradle of true art and true science. Whoever does not know it and can no longer wonder, no longer marvel, is as good as dead, and his eyes are dimmed. It was the experience of mystery – even if mixed with fear – that engendered religion. A knowledge of the existence of something we cannot penetrate, our perception of the profoundest reason and the most radiant beauty, which only in their

primitive forms are accessible to our minds – it is this knowledge and this emotion that constitute true religiosity; in this sense, and this alone, I am a deeply religious man.

[Gleiser]

Einstein called this religious inspiration for science "cosmic religious feeling". He referred to it as the "strongest and noblest motive for scientific research", a fruit of a "deep conviction of the rationality of the Universe", and finding expression in a "rapturous amazement at the harmony of natural law".

Chapter 11

Thinking of EGGS (or DEGGS)

Some readers might find what follows to be outside their normal sphere of thought but they should not dismiss it as technical gobbledegook. It does not require great theoretical know-how if basic facts are accepted, but it does require perception. It is here that one's ability to think in pictures will be put to the test. This chapter is an important feature of this hypothesis, forming one of the foundation stones of the basis of everything.

In chapter 5, I gave a very brief résumé of how the universe was formed as a result of the Big-Bang. It was perhaps too much to expect the reader to accept, without challenge, that material could be created from pure energy. To understand of what, material actually consists we need to look more closely at its formation from electromagnetic waves.

Nowadays readers have a general awareness that all material is made up of molecules and that, in turn, molecules are made up of specific combinations of atoms. Most people accept that atoms consist of a nucleus (which comprises a number of protons and neutrons) which is surrounded by electrons. The big question is; of what are *they* made?

The atom pictured as a miniature version of the solar system is the usual model of the microscopic world promoted by teachers who perhaps find they are lacking a complete understanding in this area. This is not intended as a slight on teachers; it is recognition that nuclear physics is a most specialised subject. However, as a non-scientist but yet someone who thinks about fundamental matters, I have built up a mental picture of the structure of matter - as well as of life itself. I don't have the mathematical ability, nor do I have the experimental facility to attempt to support what follows; I just entreat those that do to read my philosophical deliberations seriously. It is just conceivable that a clue to new avenues of scientific consideration will be discovered which could lead to genuine progress.

Man's understanding of the structure of the atom has undergone many reappraisals as new theories and new experimental results have gone further and further down the road of reductionism. While intensive efforts are being made to find a formula bringing together a Grand Unified Theory (GUT), we seem to be going further away with the 'introduction' of ever more constituent parts to the equation of the atom, even to the level where components only exist as a mathematical probability.

So let us start from the outside of the atom - where the electrons orbit the nucleus, not like planets in the solar system but as a spherical 'cloud', with layers at differing radii. One of the problems to be addressed is the wave/particle paradox of the characteristics of the electron; sometimes it appears to be a particle (as one might expect) but, at other times it reacts like a wave.

Sir Isaac Newton, probably the most well known scientist of the renaissance period, was of the opinion that light had a 'corpuscular' nature whereas, in contrast, Christian Huygens, the celebrated Dutch mathematician and astronomer, developed the wave theory of light. Is it possible that both are, to some degree, correct?

If the electron *is* a particle we would need to delve further into *its* structure to find out of what *it* is made, and so on ad infinitum. So let us, firstly, consider the possibility that particles do not exist at all and that the electron is a wave.

Before proceeding further, we might reflect on the probable nature of this wave. It appears likely that it would be an electromagnetic wave - stemming from the burst of pure energy at the creation. I suggest, however, that it has a gravitational element. This would be related to the curvature of the charge in the general spherical form taken around the nucleus. We will consider this in greater depth as we proceed but, as an assumption, I propose to label them as Electro-Magnetic-Gravity (EMG) waves.

If an electron is an EMG wave surrounding a nucleus, the smallest 'orbital' radius would have to be consistent with a circumference of a single wavelength since it is not possible to have a fraction of a wave without a manipulative circuit. This, I believe, is not an orbit but a travelling spherical wave, forming a vibrating shell somewhat similar to a disturbed soap bubble. The very stability of this minimum shell with a circumference of one wavelength prevents it being drawn in to the nucleus. An electron is, therefore, effectively a quantum of energy comprising one wavelength of the EMG Wave.

Since a single wavelength, turned back on itself to form a plane circle - or more properly the vibrating axis of a sphere - would have two nodes (where the value of the charge is zero), another such axis of vibration can lock in at right angles, intersecting at the same two nodes. This then allows two electrons - but no more - at the same minimum radius, as shown by Diagram 7 below.

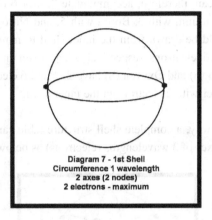

Diagram 7 - 1st Shell
Circumference 1 wavelength
2 axes (2 nodes)
2 electrons - maximum

The element hydrogen has only one proton, balanced by one electron. Helium has two protons (and two neutrons) balanced by two electrons. Elements having more than two protons will have additional electron shells which have to be formed at greater radii. The next radius available would be consistent with a circumference of two wave lengths. This being the case, four nodes will be formed allowing the intersection of three such double waves. This is shown by Diagram 8 below.

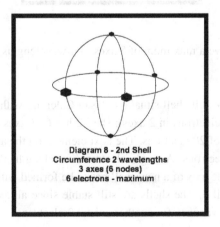

Diagram 8 - 2nd Shell
Circumference 2 wavelengths
3 axes (6 nodes)
6 electrons - maximum

99

This second shell can, therefore, accommodate 2, 4 or 6 electrons. If, as in the case of the element Lithium with 3, Boron with 5, and Nitrogen with 7 electrons, the odd electron would be drawn from the inner shell to maintain a stable second shell with 2, 4 and 6 electrons respectively. The even numbered elements of Beryllium (4), Carbon (6) and Oxygen (8) have 2, 4 and 6 electrons respectively in the outer shell, together with 2 electrons in the inner shell.

Proceeding in this way, a complete shell structure table can be built up; the 3rd shell having up to 4 axes of 3 wavelengths (electrons) as below;

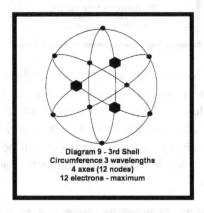

Diagram 9 - 3rd Shell
Circumference 3 wavelengths
4 axes (12 nodes)
12 electrons - maximum

A 4th shell would have a maximum of 5 axes of 4 wavelengths; and so on.

If we now look at the full shell situation of say Calcium, with 20 protons (Z=20); shell 1 has 2 (single electrons in 2 axes); shell 2 has 6 (3 axes of 2); shell 3 has 12 (4 axes of 3); a total of 20 electrons. The next element on the atomic number scale, Scandium, has 21 electrons. A single electron (wavelength) is unable to circulate outside shell three; the axis of a new shell has to be formed with 4 electrons, taking 3 electrons from shell 3. The shells are still stable since all axes are full. The 3rd shell, however, has a vacancy for another axis of 3 electrons.

100

Any total number can be made up this way; any axis always having the appropriate number of electrons to form a circumference. There is never more than one axis vacant within the system of shells - as shown in the series of tables appended on pages 186 and 187.

> The explanation given above for the shell structure seems to answer Max Planck's mystery of discrete energy levels, or quanta, in atomic structure. Although, c1900, he was the 'inventor' of Quantum theory due to his investigations into blackbody radiation, he was unable to explain, by classical physics, why these discrete levels existed. Whatever, my limited understanding of quantum mechanics is based on the belief that a quantum is simply one wavelength of EMG energy at a frequency appropriate to its setting in the whole spectrum of energy levels.

Examination of the table in appendix A shows that the so-called Magic Numbers - where greater stability occurs - coincide with the total number of electrons necessary to completely fill shell 1 (Magic Number 2); shells 1 & 2 (Magic Number 8) and shells 1,2 & 3 (Magic Number 20). However, Magic Number 28 applies to when just 2 axes of shell 4 are added, Magic Number 50 when 2 axes of shell 5 are added and Magic Number 82 when 2 axes of shell 6 are added. It is postulated that the larger shells - with many axes of vibration - are becoming increasingly unstable. However, two axes - set at right-angles to each other - do introduce a stabilising structure to the outer shell.

If this principle is accepted, that the electron is normally a wave circulating around a nucleus, from where is the particle effect derived? The table indicates the stable arrangement of the electron cloud around the atomic nucleus of each of the elements. If, however, an electron is physically knocked out of a shell, the single wavelength could form a sphere independent of a nucleus. This 'ball of energy' will assume inertia and act as a particle having mass as given by $M=E/C^2$; the pure energy having been converted into mass according to the Einstein formula.

A free electron is then a spherical wave having a circumference of one wavelength - an Electromagnetic-Gravity-Globule - an 'EGG'. These free spherical waves - EGGs - appear to be what is commonly referred to as particles. This then explains why, sometimes, they have the characteristics of a wave (because they are a free spherical wave) while they also act as a particle because the pure energy has been converted into mass. Presuming the whole of the foregoing to have a basis of truth, it is postulated that all matter is a combination of waves which, in their free quantum form, are EGGs (Electromagnetic-Gravity-Globules). We might even call them DEGGS – Duffill's eggs, to differentiate from those of the chicken, or the Cosmic Egg dreamt up by Lemaitre in the early 20th century, the exploding of which subsequently was scathingly called the Big-Bang.

Let us now consider from where the gravity component is derived.

An alternating electric charge generates an alternating magnetic force at right angles to the charge. It is possible that gravity is the vector sum of the instantaneous electric charge and the instantaneous magnetic component. Since all 'particles' (or DEGGS) are made up from electromagnetic waves curved in on themselves, this combined force would be centred on the sphere - the centre of gravity. Antiparticles (with waves travelling in the opposite direction to particles) would have their gravity vector directed outward - creating anti-gravity. The existence of some antiparticles could account for Einstein's mysterious repulsive force which, in turn, influences the expansion rate of the universe.

The alternative to the above is that the electron knocked out of the atom does not form a sphere but remains a section of wave spiralling away. This is a photon, or pulse, of light - not having any gravitational component.

It is known that unlike 'particles', or DEGGS, of matter, photons do not have mass. This is explained by geometry; a DEGG being of spherical form is closed geometrically – that is a line on the surface can be continuous; without end. It is in this way that the electromagnetic resultant forms that which we know as gravity. This inertial gravity is equivalent to mass. On the other hand, a photon of light is structurally/geometrically open. The waves of the packet of light are radiating out from a source. Not having a closed geometrical structure, gravity does not result – there is no mass. However, since a photon is an electromagnetic wave, it can be influenced by the gravitational force of a large mass; hence distortions do occur in light being received from beyond any star. This phenomenon also accounts for the existence of black holes, where intense gravitational forces draw light into the core and prevent the radiation outward of any light.

The energy of a wave is related to the Root-Mean-Square (RMS) value of its parameters. Traditionally, when taking the square root of the mean of the square of all the instantaneous values, the resulting possible negative value is rejected as being irrelevant in favour of the positive value. I believe it depends in which direction the wave is travelling. Considering electromagnetic waves, travelling in one direction the positive root would be appropriate to a positron. Travelling in the opposite direction the negative root would be appropriate to an electron. Hence, if the two meet *and the nodes coincide*, they annihilate each other.

We can extend the whole principle of spherical energy waves to the nucleus of the atom. It appears probable that protons and neutrons are formed in a similar manner to electrons but with much shorter wavelengths/higher frequencies giving greater penetration at higher energy levels and having proportionally greater mass. It is also likely the nodes of the neutrons link with the nodes of protons, thereby providing the binding forces needed to prevent disintegration. Quarks and other unstable, so called fundamental 'particles' may simply be harmonics which, when disturbed from their primary waves, form minute short-lived DEGGs.

It is interesting to note that the attraction force between quarks increases in proportion to the distance between them – like as if the are connected by elastic bands; the more they are stretched, the stronger the force. It is probably this fact that caused them to be drawn together in the primordial soup to form protons and neutrons.

I must admit to lacking a full understanding of the sub-atomic world of Baryons, Mesons and Leptons together with their associated Gluons, Pions, Kaons and the many other minute constituents; it is a very complex sphere of research, I suspect not yet fully understood by the specialists. Regardless of my inadequacy, the point being made here is that the whole of an atom appears to be comprised of waves. This being the case, then **all matter is comprised of waves** - even the smallest of 'particles', since they are themselves, spherical waves.

Let us look more closely at the concept that the elusive force, gravity, is the vector resultant of the electric charge and the associated magnetic force. Since the component parts of the atom: the electron, the proton and the neutron, are spherical in nature, the combined resultant of their charges and magnetic forces are directed to the centre of each component. It surely is this that 'creates' the mass from the pure energy of the originating waves.

Large masses, such as galaxies, will have sufficient accumulated electromagnetic gravity, to significantly deform Creation's primary waves, causing a distortion of the time lines - warping space itself.

We started this cogitation by considering the probable wave structure of the electron and the constituent elements of the nucleus and then expanded briefly to the galaxy. It is suggested here that the link between the domain of quantum physics and that of the cosmologist is the DEGG (the Duffill-Electromagnetic-Gravity-Globule) and, possibly, the anti-DEGG.

Similarly, on the cosmic scale, the whole universe is composed of electromagnetic waves; these will be curved - on the grand scale - centred on the point of Creation, the Big-Bang, creating the mass of the whole universe. This being the case, total gravity is not only dependent on the amount of material in the universe but also on the potentially convertible energy. I believe that, if the calculations are processed from that hypothesis, sufficient mass - or potential mass - will be found to cause the expansion of the universe to slow to a state of momentary equilibrium before commencing an ever increasing contraction.

As the contraction gains momentum, temperatures and pressures will increase, converting all matter back to pure energy. When the dimensions of the universe have reduced to infinitesimal proportions, the birth of the next Big-Bang will occur - as the inverse of the current one. The completion of the grandest possible half wave will have occurred; this to be followed by a matching negative version – where positrons dominate, in lieu of electrons. The whole creation - as with everything within - is simply a wave.

We see from Diagram 10 (next page) that the Big-Bang does not suddenly occur; it is the inevitable result of the accelerating shrinkage of the universe to a singularity – a chaotic nothingness where all rules are broken. As the 'particles'; the electrons, the photons, and the quarks, shrink towards infinite density, mass is converted to energy travelling close to the speed of light where time stands still. Although this is deep into Heisenburg's uncertain quantum mechanical world it is also still in Einstein's world of space curvature and the ultimate black hole. The expansion that necessarily follows the ensuing cataclysmic explosion will be into the inverse half-cycle to repeat the whole process, probably with inverse polarities.

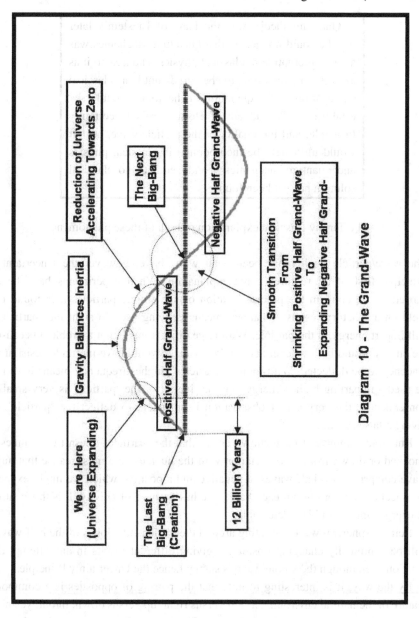

Diagram 10 - The Grand-Wave

> Quantum theory was the bane of Einstein's later life; he could not accept that Quantum Mechanics was a valid alternative to classical physics – he treated it as an interim measure to be used until a classical explanation of quantum behaviour could be established. The advent of Heisenberg's Uncertainty Principle, and the duality of the particle/wave, he felt would ultimately be incorporated in classical physics and quantum mechanics abandoned; but no classical solution has yet been found.

I offer my classical explanation of both of these phenomena.

The wave/particle duality has been dealt with above; however the Uncertainty Principle first of all requires some explanation. When experiments have been carried out to confirm the precise location of sub-atomic particles it is found to yield unsatisfactory results. Whatever means is being used to sight the 'particle' will impart energy to the particle. With large objects this is not a problem because the energy would be low level and the larger mass would have inertia to resist any tendency to be deflected. With minute 'particles', higher-frequency means have to be used - imparting higher energy, and the inertia of the 'particle' is very small; consequently, the very act of observation is sufficient to deflect the 'particle' – giving a false result.

But, wait a minute, I have shown above that the 'particle' doesn't exist unless knocked or drawn from its association with the nucleus. We must realise that any wave comprises two half waves; a negative and a positive which we might expect to cancel each other out but each half wave has a centre of gravity. It is this centre of energy which would be detectable.

Being a spherical wave, vibrating around the nucleus, the lobes of the half wave will be continually changing position – giving different results in any attempt at location even though the whole wave is stable; hence the Uncertainty Principle.

By the way, it is interesting to note that the pairing of opposites is a common feature of the natural environment; left versus right, up versus down, in/out, yes/no, on/off, rich/poor, male/female, and so on. There are many opposites that do not cancel each other out; they are relatively complementary.

These then are *my* thoughts. Whereas they may not agree with established current quantum theories, they do give a plausible explanation to the structure of matter. Scientists should view this hypothesis, not as an opposition to current thinking, but as complimentary – a framework of alternative thinking within which the established details will find a fit common to all disciplines.

Linked with previously stated thoughts related to the philosophy of life, waves do seem to be the link between everything – in the cosmological world, the quantum world, and the everyday world of our existence.

All matter and all manners are formed by waves.

Chapter 12

Thinking of Harmony

We are all aware that the universe as a whole is highly complex; even the complexities have complexities, and so on ad infinitum; how can we begin to understand what makes it tick? How is it that the many facets of this world in which we exist seem to fit together whether they are large or small; solid, liquid, or gaseous; organic or inorganic; every lasting feature being compatible - at least indirectly, with the whole.

In order to investigate the cause of something it is common practice to examine relevant details ever more closely; we make use of electron microscopes to witness cellular activity, or cyclotrons to explore the world of sub-atomic particles. We probe the human body with X-rays and powerful magnetic beams to discover the innermost workings. These reductionist techniques educate us regarding specific attributes but they lose sight of the whole. A suitable analogy can be found in the world of crime detection. The laboratory which establishes that the DNA of a sample of tissue or a sample of blood belongs to a Mr Smith does not show the said Mr Smith is guilty of some crime; the technician who is probably unaware of the overall circumstances is not in a position to judge. To make a case, the prosecutor has to show how the details link with all other information in order to create a whole picture.

Somehow we need to show how the myriad of cosmic details fit together to form the multi dimensional jigsaw we call the universe.

The preceding pages have indicated how all matter, whether living or not, can be reduced to electromagnetic waves; the workings of the brain depend on wave transmissions both internally and externally. EMG waves from a variety of sources can combine to good effect if they are in harmony. We must remember that details have to be related harmoniously within a sector of the brain to create a stable characteristic. This compound characteristic then must be related harmoniously to other compound characteristics in other brain segments and so on until the complete personality is established. In an earlier chapter, I suggested the analogy that the documents in a file need to be related; the files in a folder need to be related; folders in a drawer need to be related; and drawers in a cabinet need to be related. We then have a complex amalgam of details which lock together in a unique manner.

A personality comprises many sub-characteristics harmoniously combining to form the whole; they must be in harmony if the end result is to be a stable character. This complete personality characteristic is mirrored by an existing wave combination permeated by the equally complex, multidimensional pattern of ethereal waves stemming from the Creation. As a result, the complex personality characteristic wave harmonizes with a particular combination of ethereal waves. The harmonious match that is established is the mind/soul link.

This hypothesis is not as unreasonable as one may first think; if you visualise the primary waves radiating from a source – the Creation; the Big-Bang, the presence of celestial bodies will obviously be interference. These bodies absorb the energy received and re-radiate it in a modified form. The multitude of sources of re-radiated energy will create a very complex pattern of wave interference. At any given point in the universe waves from many sources will combine to make a complex characteristic in a similar manner as the developed characteristic in an individual's brain.

It depends greatly on the developed characteristic as to whether the link aligns predominately with the spurious waves from the many sources within the cosmos or whether the main influence is that of the pure energy waves emanating from the creative Big-Bang. If the former is the case one would expect a looser character than the more disciplined character built on the latter.

As suggested earlier, characteristic waves develop as life presents new experiences; conflicting waves being rejected in favour of those more compatible. If new experiences are strong enough to cause a change in the overall personal characteristic, the mind/soul link will adjust to a new harmonious juxtaposition of ethereal waves. Very traumatic experiences such as, for example, the witnessing by a child of a violent, bloody attack upon a much loved parent can *never* be totally expunged no matter how diluted the event may become with time; this is the principle of homeopathy, which is discussed in chapter 14..

Other personalities may well have developed very similar characteristics – either currently or in the past, and will, therefore, have established a link very closely related to the same juxtaposition. This *is* the harmonious closeness referred to between family members and, of course true lovers; but it is also the possible explanation why a living person has an affinity with someone from the past; a talented musician perhaps.

When we dream, it appears that our subconscious is allowed free reign; the discipline of our characteristic is relaxed. As a result, our mind/soul link wanders to other ethereal characteristics giving us a picture of other 'worlds'. Since electromagnetic waves travel at the speed of light, for them time stands still. These other worlds can, therefore, be forwards or backwards in time; premonitions and trips into the past can be understood in this context. The mind/soul link can also wander sideways, perhaps led by unfulfilled desires or fantasies, giving insight into the characteristic domain of other personalities. These other personalities may not yet be linked to any actual person past or present, but could be purely figments of the imagination. Such excursions will probably be triggered by recent experience, no matter how obscure. We must remember that the new waves of experience, if found to be harmonious to a sector of the mind, will cause a modification to that sub-characteristic triggering a drift in the accumulated personality code.

When we are not asleep and not in some form of trance, our general characteristic is likely to resist any tendency to wander; we suppress innermost desires to mentally explore exotic domains. This means we are slightly false but this is a necessary restraint to avoid aimless mental wanderings. An undeveloped personality does not have this strong fundamental characteristic capable of absorbing new experience without significant distortion and therefore displays instability resulting from uncontrolled mental wanderings. The ingesting of drugs – whether by prescription or otherwise, is the probable cause of similar effects culminating – in extreme cases, in experiencing hallucinations.

> On page 97 of *Unfinished Man* by Raymond Van Over, Carl Jung's deliberations relating to dreams are reported. The celebrated Swiss psychologist felt that there are two fundamental points to be kept in mind when dealing with dreams: First, the dream should be treated as a fact, about which there should be no preconceived assumptions except that it "somehow makes sense"; and second, that the dream is a "specific expression of the unconscious." The unconscious, to

Jung, expresses a truer and more complete picture of what we are. A free communication between the conscious and the unconscious creates a healthy communion between an inner reality and the external world. Dreams are an important method of communication between the two worlds.

[The hypothesis presented in these pages provides evidence of a physical link between different 'mind-worlds' – the characteristic electromagnetic waves and the mind/soul link.]

The fact that electromagnetic waves travel at the speed of light causing apparent timelessness can lead to premonitions. If our personality characteristic was tuned exclusively to the pure energy waves from the Creation (the All-Enveloping-Influence) we would, no doubt, have a clear picture of what lies in store; this would be true communion with God. However, due to the complexities of the intermingling of cosmic waves, the picture received is unclear and probably not recognized as a 'glance into the future' until the event actually occurs.

I have had two quite remarkable personal incidents which appear to illustrate this possibility.

In the first incident, as a child of 9 or 10 years of age, at the early stages of WWII, I was very knowledgeable regarding the many types of aircraft currently in use, both by the enemy and the allies. I, like many a youngster, fantasized about flying; I wanted to be a pilot when I grew up. I was so obsessed that, while lulling myself off to sleep, I invented a simple aeroplane for me to fly. It consisted of straight wings mounted on a regular fuselage having a single fin and rudder. On the wings were two hollow cylinders in which were mounted a series of rocket-type fireworks; by firing them in sequence they provided the necessary thrust to cause the aircraft to zoom off into the sky. Nothing strange about that you might conclude unless one appreciates that this was in 1940, long before the propeller-less, jet aircraft appeared to the public eye. The remarkable aspect of this story is that in the 1950s I became a jet pilot mainly flying the Meteor which was an aircraft with straight wings embodying two hollow cylinders in which were mounted the jet engines to provide the necessary thrust.

The second unintentional apparent forward projection of my mind occurred in the mid 1980s. I started to write a novel whose leading characters were Sam and Sue Jackson. He was an Engineering Projects Manager in England's Yorkshire undergoing a period of dullness due to being in a career groove. In the story, Sue notices an advertisement in the newspaper for an Engineering Manager for a demanding project in the Scottish Highlands; no details were given other than a box number for any replies. Sam applies and receives an unexplained telephone call giving arrangements for a flight up to the Highlands where he and Sue will be met. Somewhat apprehensively, they embark on this mystery trip. After discovering what was on offer, Sam takes on the challenge and the story proceeds. In December 1988, when the story was 70% written, my wife showed me an advertisement in the newspaper for a career opportunity in the Highlands of Malaysia. Although I doubted my suitability for the position, I applied. The result was a strange telephone call - after business hours on a Friday, requesting that we both travel to London for a preliminary interview. This done we then received a further telephone call with arrangements to fly out from the UK to Singapore where we would be met and then flown up to the Highlands of Malaysia; there was nothing in writing, we didn't even know the name of the company who were our potential employers. Resulting from taking this quite dramatic trip, we found the offer to be agreeable and accepted the offer. The 'coincidence' (or was it a 'glance into the future') was that we started a new career in a similar manner to the fictional Sam and Sue, flying off to an unknown destination – in a highland situation, without knowing who was calling the tune. [The full story of our most unusual adventure forms part of my autobiography, "Maurice Duffill's View of Life", and the novel is entitled "Highland Haven".]

These are examples of possible glances into the future; but are the mind/soul links the possible explanation of clairvoyants? Given that the medium, a person with heightened sensibility, has been provided with an insight into the characteristic of a subject, he/she may be able to tune to the appropriate mind/soul link in order to discover the relevant past or future. This would appear to be more probable if the subject was involved in some emotional event – a violent death perhaps, giving off strong vibrations. I have an open mind on this sphere of interest.

Extra Sensory Perception (ESP) has been the subject of much controversial research without producing conclusive proof of its true behaviour. Much of the difficulty has been due to the history of charlatans and the difficulty in arranging truly objective tests. I suggest that the transmission by the brain of electromagnetic waves, as discussed in these pages *does* have the ability to be received by others – but there must be an emotional link, such as that between a truly-loving couple, in order to separate the subject waves from the general clutter - the noise created by the plethora of ethereal waves. Again, I have an open mind on this matter and believe one should not dismiss that which cannot yet be proven; one day a fuller understanding will become available.

The opposite of harmony is, of course, conflict. Conflict occurs when the characteristic waves being received are not in phase with the established personality. If the received waves are strong and the personality weak, distortion will result and the personality is further weakened. On the other hand, if the received waves are compatible, a strong personality will, after a minor adjustment, absorb this new information/experience; increasing the breadth of character.

It is simply a matter of do the waves fit or not?

Chapter 13

Thinking of God

In the preceding pages, I have expounded on the theory that the root of *everything* is the electromagnetic-gravity wave. I have shown how *all* matter progressively develops from the pure energy of these waves and that from the basic elements, created in the intense heat and pressure of the active cosmos, an endless variety of molecules have evolved. This variety of molecules forms all the materials to be found, not only on Earth but in the entire universe. Certain molecules have united into living cells which, over a period of time, have developed into complete bodies, both vegetable and animal. The 'magic' of evolution has led to a phenomenal spectrum of plants and an incredible range of animals - culminating in us, the so-called intelligent, human beings.

It does take some considerable thought to accept that all 'things' have evolved from nothing more than pure energy in the form of electromagnetic-gravity waves. If, however, one goes back and forth through the preceding text, the soundness of this hypothesis begins to become clear. Any material, whether living or dead, must be made of something; but then that something must be made of something, and so on. This progression is only true to a point. We reduce material down to molecules; molecules reduce to atoms; atoms are made up of their components, protons, neutrons and electrons. Protons and neutrons are comprised of quarks. Here the point has been reached when we say "Wait a minute, quarks and other sub atomic, so-called 'particles' do not have a stable existence as matter"; a quark cannot be held in readiness for use in building more complex matter.

Quarks are like sparks; they are purely transient pulses of energy, capable of causing a reaction. I suspect quarks are tiny harmonics radiating from the primary electromagnetic-gravity waves emanating from the Big-Bang. Given a suitable combination, these minute harmonic waves (probably in an unstable spherical form) amalgamate in the intense heat and pressure of a supernova or the Big-Bang to form more durable matter such as protons or neutrons, but comprising nothing more than waves of energy. I think we will leave the details to the nuclear physicists but there appears to be no doubt that the only way that pure energy can be converted into matter is by the re-arrangement of the primary electromagnetic-gravity waves and their harmonic components.

From the preceding declaration, we come to realize that the moment of Creation must have been preceded by the moment of conception. If, at that moment **nothing** existed bar pure energy, then God must be pure energy. This apparently outrageous conclusion will be found to fit all the requirements of The All-Enveloping-Influence, the spiritual overlord of all Creation.

My definition of God
God is the pure energy, in the form of electromagnetic-gravity waves,
from which *all* matter is created.

One is, of course, aware that the majority of people visualize God as a man - frequently referred to as He (for some, apparently sexist reason). This mental concept presumably stems from the reversal of the teaching that man was made in God's image - therefore God must be like man. It is very presumptuous and very naïve to believe that God looks like *us*. I believe we *are* made in God's image because, only because, in the ultimate reduction, we are created from the pure energy which *is* what we choose to call God.

Not content to be made from waves, we humans respond to waves by virtue of our inbuilt 'radio' tuner. Our personal characteristics are matched by a combination of ethereal waves. This combination, as discussed earlier, is what we refer to as our soul.

Let's again visualize the 'ether' cluttered with radiations (just like radio signals) coming from their billions of sources in every conceivable direction. We switch on our brain and scan the noise - just like searching the short waveband for stations of interest. Faint responses are detected from unknown sources but then, as we scan, stronger messages are received from more obvious regions.

Some of these clear messages we appreciate and perhaps spend time absorbing the content - since they come from close family, friends, or lovers. Then again, some we reject and switch away from - they come from incompatible associates. Through all the clutter there is the one frequency which can always be heard; that of the Principal Ethereal Vibrations (PEVs) which have been there since the beginning of time.

PEVs act as broad band transmissions such that one does not really need to search; these particular waves appear all around us, all the time. It is only necessary to de-tune from the other incoming messages to be able to absorb the All-Enveloping-Influence; the signals from the Creation; God's messages.

This is the purpose of prayer; to focus - consciously or subconsciously, on the Principle Ethereal Vibrations giving access to the unadulterated source. The All-Enveloping-Influence offers the necessary support to those individual minds able to recognize it, by essentially being the stable base for everything.

Some scientists pour scorn on the ability of God to be everywhere all the time and his ability to give support to individuals. The concept of ethereal vibrations provides the basis upon which everybody and everything – throughout the universe – not just Earth, has a common connection to the creation – hence the availability to the All-Enveloping-Influence.

> Richard Dawkins reports in *The God Delusion* that Steven Weinberg believes "Some people have views of God that are so broad and flexible that it is inevitable that they will find God wherever they look for him." One hears it said that 'God is the ultimate' or God is our better nature' or 'God is the universe'. Of course, like any other word, the word 'God' can be given any meaning we like. If you want to say that 'God is energy', then you can find God in a lump of coal."
>
> [Although he is being cynical, he has hit the nail on the head. My definition of the word 'God' clearly states the infinite breadth of influence based, not on some mythical teaching, but on the very plausible hypothesis presented here. Yes, God *can* be found in a piece of coal.]

Although emanations direct from the Creation are all around, and within us, and readily accessible, so are the multitude of radiations from the billions of sources scattered throughout the cosmos. All random transmissions are not evil, many emanate from those close or those with similar interests and aims. It is, however, necessary to be selective. This is where the foundations of one's sympathies are put to the test.

As I see it, what we commonly refer to as 'Lightness' (good) versus 'Darkness' (evil) is the acceptance of the guiding aura of Principal Ethereal Vibrations versus rejection in favour of a variety of alternatives. To reject the basis of common purpose rejects cohesion; without cohesion, life falls apart.

You may well have noticed that, although the foregoing supports the belief in God, no particular religious doctrine has been promoted; I have merely indicated my concept of how one can communicate with our Creator. It should not matter what faith (*if any*) you chose to follow; the end result should be to seek communion (a small 'c') with our fellow 'man'. Only through common purpose can this be achieved - we should have **consideration for others**.

It is not everyone who is prepared, or who is able, to apply fundamental thought to spiritual matters. For those that cannot, a host of religions provide doctrinal procedures which can give leadership and support for their belief in God. Whether a person attends a church, temple, synagogue, mosque or other place of worship usually depends on cultural upbringing but the ultimate objective is to commune with one's creator and fellow man. While receiving the ministrations or praying, the congregation are all of like mind - transmitting individually weak, but collectively strong harmonious signals. This focuses the collective body on to the appropriate frequency as directed by the minister. It can be likened to being guided to a particular pathway; in this case, the pathway should be to the Creator.

I am not a Muslim but I observe that the Hajj – the annual gathering of Islamic followers of many nationalities from all over the world, represents the ultimate in collective assembly with the coming together in very close physical proximity of up to two million people of like mind. Those present report an apparent cleansing of the soul which, of course, strengthens their belief. There is no doubt that this comes about due to the resonance of similar thought waves between minds having a common focus. Perhaps other sections of society could learn from this.

The down-side to collective assembly is that each person is influenced by those in close proximity; it is difficult to think independently when all around are giving off mental vibrations. For better or for worse; those leaders with evil intent can make use of the resonance of collective thought by directing their followers to other targets, with devilish aims. As has been discussed earlier, it is dependent on one's fundamental characteristic to recognize and reject undesirable influences. Some people - and I *am* one of them - prefer to be isolated from the distracting thoughts of others so that the mind is free to travel in any chosen direction. This way, one can more deeply pursue avenues of thought without being forced down established doctrinal channels.

In times of personal stress, people react differently. Some will seek counselling or the comfort of sympathetic companions to focus their thought waves to an acceptable frequency or wavelength - to build on the stability of the Principal Ethereal Vibration. Those more used to free thought will probably seek solitude, or the support of only the closest of companions, to achieve the same objective; whatever the case, positive effort *is* required to successfully direct the mind in order to deal with the crisis. Without purpose, the mind will be vulnerable to undesirable influences from the many signals being received. Some of these will be from within such as anger, envy, even thoughts of revenge and those from without which are socially unacceptable, perhaps of criminal substance.

In such times of great personal stress, how can one sweep aside undesirable or irrelevant thoughts? A technique that my wife and I have adopted when faced with a significant turning point in life (of which there have been quite a few) is to 'start from here'. Put behind all the thoughts and reasons as to why one is in the current situation, simply consider the options available. This way, emotions based on past events will be quashed and all energies directed forward; this, in turn, leads to clearer thinking. *We have applied this procedure quite successfully after loss of income due to a serious accident; after being ejected from our country of residence due to politics; and after losing our home in a bush-fire. It has led to a definite improvement in our quality and appreciation of life.*

My short sermon is now over. I apologize for that diversion but it serves to illustrate how the waveform of one's characteristic can be found to fit new associations - for good or evil - if the distractions of waves of conflict can be set aside. I mention the alternative of evil because the same process is applied by people engaging in unacceptable behaviour; they set aside their thought waves which are based on consideration for fellow man. The current concern regarding terrorism has developed largely due to this technique of progressively and

118

collectively tuning the minds of susceptible groups to the point of being prepared – in fact 'honoured' to give their lives for the cause. We saw this at the latter stage of WW11 with the advent of the Kamikaze pilots.

At the risk of repeating myself, I remind you that wave combinations in the multiple dimensions of time and space, found in the cosmos, can - under conditions of great heat and pressure - form into highly complex combinations resulting in matter. This matter itself is, in some cases, then further influenced by combinations of waves to become living matter - a biological being. Some biological beings - those with a mind - are even further influenced by waves to develop character. That character has had all his/her experience encrypted into a complex waveform which is suspended from inevitable decay by the reception of waves from current sources which strengthen or modify the developed waveform of the mind.

It is not surprising that this confusion of waves creates a cloud obscuring fundamental thinking. It requires positive effort to penetrate this fog in order to reach the fundamental truth.

We've exercised our brain throughout the preceding chapters, exploring thoughts on many aspects of life - but what about death?

There is much controversy regarding the possibility of an after-life, in fact many religions consider this to be the whole point of life on Earth. It is often suggested that the soul rises from the body to remain forever in the celestial cosmos. According to the hypothesis presented here, we have a measure of agreement with that vague idea. As discussed earlier, the soul is the guiding characteristic which programmes the mind and can be found at the juxtaposition of appropriate ethereal waves. In other words, the four dimensional convergence of all these waves, modified by particular harmonics which make you *you* is, I believe, your soul. This abstract psyche is the sum of all the major influences from which the character - the person - has developed.

Even when that person ceases to exist as a being, that psyche is still formed - it is not dependent on the presence or otherwise of a body. The point in the space-time environment of the celestial cosmos where this convergence occurs could well be that which is often referred to as 'heaven'. It is worth noting that, mathematically a point has position but no magnitude; it therefore follows that heaven will never fill up. There is already an infinity of such convergences waiting for a developing being to tune in and then lock on to that combination of frequencies.

Some readers may find this extension to my concept rather difficult to accede to but it appears to satisfy all the 'requirements' of the soul - from birth to death - and beyond. This being the case, we should have no fear of death; one's characteristic (usually referred to as 'spirit' after the body has died) continues on without the encumbrance of a body. Loved ones need not feel sorry for the departed, only for their personal, temporary loss; their spirits (if unchanged) will come together in due course.

Although this last statement is a common belief, we can now see how – in the context of this hypothesis – this actually can happen; both spirits, having very similar characteristics, will be brought together by their fit to the particular combination of ethereal waves. Believing this to be true can be of great comfort in times of bereavement.

Understanding that the love shared between individuals is related to the closeness of their souls - rather than their living bodies, is also very relevant when considering the controversy regarding organ transplants. Since these body parts have no influence on the developed soul, no love will be lost by allowing the use of these organs to save other persons lives. For the same reason, the apparent mutilation carried out during an autopsy will have no effect whatsoever on the characteristic developed during life and now secured in the equivalent juxtaposition of ethereal waves we call the soul. The body is a transient vehicle while the soul lives on.

Part 3 – The Review

Chapter 14

Thinking Back

Do electromagnetic waves provide the link between everything - either physical or metaphysical? I think the preceding chapters amply illustrate that they do. The basis of all material - whether inanimate or animate - is shown to be the electromagnetic-gravity waves emanating from the creation of the universe - the Big-Bang. We have considered the material aspects of Creation's waves as well as the psychological influences of the many sources of electromagnetic radiations. But what makes man human?

The principal difference between man and animal is language; not just the language of speech but the language of art, of music and of mathematics. Animals display little evidence of these attributes although birds do sing and some animals appear to be capable of being trained towards counting and speech (although the evidence is somewhat flimsy and controversial). Mathematics provides the logic that we humans display with regard to problem solving; not just the manipulation of figures but the ordering of tasks and decision making. Music and other art forms represent the harmony of the thought processes which can calm or stimulate higher levels of awareness and excitement depending on one's choice. Speech is the essential means of communication, not necessarily via the mouth but by the written word or symbolic directions. More importantly, speaking internally, within the mind, is the means by which we humans direct our thought processes. Without language, we could not think arbitrarily.

What has all this to do with waves? As discussed earlier, each of these attributes is developed (preferably from early childhood) by the superimposition of appropriate wave-forms on to the fundamental characteristic of the individual's mind. Notice how, for example, a child brought up in a musical household readily develops an awareness of the finer points, the subtleties, of this most beneficial art-form; he/she can *think* musically. While the mind is relatively free of 'clutter', a firm waveform incorporating musical attributes can be established - these harmonics of the characteristic are being added - being remembered. Even the apparently natural talent of perfect-pitch, which some people display, is a learned and remembered attribute; it is a matter of being aware that a particular sound, say 440 Hz, is that of the 'A' note, or 264 Hz the middle 'C' note. As a child you would have no idea what to call that sound unless told to label it 'A' or 'C'. The same applies to colour; 'blue' is the name given to electromagnetic vibrations of a particular wavelength (480 nm for Cadmium Blue) or 'red' for the longer wavelength of 644 nm [nm is short for nanometre which is a billionth of a metre]. We could just as easily have called them by other names or, what is red could have been named blue and vice-versa; the point is that such knowledge is learned very early in life.

A child born to an environment of mathematics or literature will enjoy equivalent benefits to those born into a musical household; the solution of equations or the structure of a sentence will develop as a natural attribute. Conversely, anyone unfortunate enough to be born into an unsettled environment - not having any stable input - will have great difficulty in ever establishing a firm, harmonious characteristic. Because of this, it is likely that such a person will be unable to deflect the unsociable traits encountered and will develop a weak character or a strength based on generally unacceptable behaviour.

It all comes down to what waves are being received when the person's mind is vulnerable - early childhood (when they are more readily remembered) or at some later critical moment in life. Born in England, I naturally learned to communicate in the English language. When in my mid 60s, I joined a Welsh choir whose repertoire included a large proportion of songs in that language. After singing Welsh for over seven years, I still have little knowledge of the meaning of the words; I have learned the pieces by rote – the language has not been assimilated into my characteristic. Had I been immersed in the language at an early phase in my childhood, the result would have been quite different.

While stressing the importance of the establishment of a memory bank of characteristics, let us consider one of science's thorny problems, 'homeopathic dilution'. Homeopathy works on the principle of providing a dilute mix of the substance causing the patient's malady or one able to create similar effects. Specialists in this field have found that, to avoid toxicity, it is necessary to dilute the substance to such a degree that not even one molecule of the introduced chemical is left in the carrier liquid (usually water or alcohol). The problem is, if not one single molecule is left, the patient is surely being offered pure water, or pure alcohol. It is for this reason that scientists have been very wary about the principles of homeopathy.

In recent years a French scientist, Dr Jacques Benveniste, believed that he had discovered that the fluid in question had somehow remembered the effect of the introduction of the substance. This was challenged by a team inspecting his procedures and the theory disclaimed, resulting in this brilliant scientist's reputation being ruined. I would suggest to you that it is quite probable that a form of memory *does* exist in these circumstances due to the 'aura' of the material's characteristic. We should not be thinking at the molecular level; let us consider the sub-atomic level. If, as I believe, the characteristic wave-form of the subject matter can be modified by the addition of other wave-forms then – to remove the effect will require an equal and opposite addition, otherwise the subject matter will carry a memory of the original addition, in the form of a distortion of the wave-form; the electrons, protons, and neutrons will be vibrating in a slightly different manner to normal.

Spectroscopy is a method used to discover the frequency of vibration of materials, showing that each and every substance does have a characteristic based on its electromagnetic wave-form. Using spectroscopic procedures a detailed analysis can be made of any material. When the original pure water/alcohol sample has the pollutant added, the characteristic wave-form of the base material is modified. Although the liquid is then repeatedly diluted, that modification still exists, albeit at an ever decreasing strength – taking a dilution of 10% repeatedly will never arrive at zero. The basis of this presumption is the fact that all material is fundamentally composed of electromagnetic-gravity waves. If another substance is successfully incorporated into a base medium, the characteristic of the latter will be modified - permanently, or at least until that added characteristic is cancelled out exactly not by dilution but by an opposing pollutant of precisely the same strength. I wonder whether or not current spectrographs are sufficiently sensitive to detect minor variations in wave-forms rather than in the frequency of vibration.

This homeopathic notion raises the thought that it is very difficult to 'clean' water that has ever had another substance truly incorporated with it. The water we drink has, therefore, been modified by many minerals to which our bodies have to become accustomed. In some areas this will lead to greater health. In other areas, the toxins would have been countered by a bodily reaction to compensate. In 1976, my wife and I visited Leningrad (now St Petersburg) and found that, as a result of the tremendous damage caused during the wartime siege, the water mains were contaminated by the sewerage system. A westerner drinking tap water or even just cleaning the teeth would suffer ill health very quickly - as I soon discovered - whereas the locals had developed immunity due, no doubt, to their constant conditioning.

The point really being made is that every small addition to a fundamental characteristic causes a change which is virtually permanent; it *is* remembered, unless an opposing characteristic is introduced to cancel it out. The same occurs with our personal characteristic; every experience, no matter how small, causes a change. If that experience is repeated, as when we are being taught, that change will be reinforced and be remembered. The prime memory is then the overlay of a new characteristic onto the fundamental. This differs from the act of remembering by rote as one does with telephone numbers, addresses and other lists of data, but even in this case, the data is normally only remembered if it is subject to frequent use.

The fundamental characteristic, or aura, of an individual does - as a result of remembering past events - truly reflect the accumulated experience of life so far. How then are these waves affecting the general philosophy of life? Let us review the questions raised in the opening paragraphs and consider whether or not they have been answered.

Is there a God?

I think there is no doubt that an All-Enveloping-Influence exists and that influence is available to all by allowing the character of the waves from Creation to be a foundation to one's characteristic. In the preceding chapter it was propounded that God is the continually existing pure energy. It is from this pure energy that all matter, including all life, evolves and all metaphysical characteristics emanate. This energy - which we choose to call God - is truly the maker of all.

Where is the soul?

The soul is the combination of characteristics of the mind which are identical to a particular juxtaposition of ethereal waves found at a point in space-time. This

124

rather ponderous sounding definition of the location is that commonly referred to as heaven. In contrast, hell is the pointless wandering of a soulless spirit. This leads to the conclusion that peace of mind will only be achieved if harmonious additions to one's characteristic are sought. 'Conflictics', such as anger and jealousy, undermine the fundamental characteristic, weakening its resolve. The weaker the characteristic, the more difficult it becomes to lock-on to any particular combination of ethereal waves.

How can God be *in* all of us?

Since the radiations from the Creation permeate the whole universe, everybody - and everything - is under the influence of the original pure energy; that which is called God. Whether we seek it or not, our personal characteristic is continually, rather subtly, receiving EMG signals from this truly central source. Whereas many other signals are received from many other sources, this central source is consistent in character. If we *'listen'* we will hear it. We can allow it to become the basis of our own characteristic or we can reject it in favour of more transient influences.

Why am I who I am?

Character is more than an intangible attribute; it is the accumulation of all the wave-forms of life's experience. Some of these wave-forms are strong, due to repeated additions of the same and harmonious experiences - particularly those from early childhood. The resulting compound characteristic is that which guides the mind - the soul. This is one's personality.

What is conscience?

Conscience is the effect created by the person's compound characteristic which determines one's moral standpoint. What is considered right or wrong is entirely dependent on the person's experience so far - the strength of the various component wave-forms which aggregate to create the compound characteristic. Conscience is, therefore, the ability to decide on a path based on one's accumulated fundamental characteristic. The mechanics of conscience raises the question: does the offered characteristic fit the existing? If the new experience does not fit an existing stable, well developed characteristic, conflict will be felt unless it is rejected.

What is meant by *self*?

Self is the awareness of one's characteristic; to know you are an individual as opposed to simply another of our species. Once one is aware of self one then has

125

the ability to advance the development of our fundamental characteristic by seeking complementary wave-forms through further education and by associating with those having a harmonious aura. The net effect is to accelerate the development in a similar manner to the working of compound interest; the more one achieves, the more one can achieve.

Until an understanding of the electromagnetic-gravity-wave characteristic of 'self' is fully developed, cyborgs (intelligent machines) will never be a reality.

What are emotions?

Emotions are strong feelings, such as love or hate, created by the similarities or otherwise of fundamental characteristics. Our love for a person is generated by a particular harmonic growing in strength. If this love is reciprocated, a resonance develops due to sympathetic 'vibes' bouncing back and forth, developing similar characteristics - creating an ever strengthening bond of common thought. Physical proximity will tend to promote an increase in the interaction of one to the other. The opposite expression of hate is similarly created by the build-up of 'conflictics', whereby thoughts of the subject-person, or activity, generate a wave-form not compatible with the person's prime characteristic. This can be exacerbated by similar thoughts from the subject-person creating a resonance of negative vibes.

Is life predetermined? - Certainly not.

We have seen how it is up to the developing individual to choose the path forward based on his/her own experience so far. The person's characteristic will govern the choice of direction depending on the relative strengths of the component wave-forms. It is for this reason that it is so important to avoid allowing 'conflictics' to deflect the prime wave-form. If the wave-forms of anger, envy, jealousy, hate, etc are added to one's base characteristic, it will become deformed; you will now be a different you - having an aura that is found less compatible with those around you. The consequence will be to seek those who *are* compatible - other angry, hateful persons. This is how anarchistic groups develop - basing their outlook on the negative aspects of life rather than the positive.

Having looked at the philosophical questions that were raised earlier, let us now consider the materialist aspect of the fundamentalism of electromagnetic-gravity waves.

How can energy become matter or vice-versa?

Einstein showed that there was a direct correlation between energy and matter with his well known, simple formula $E=MC^2$ but that does not give a mental picture of how conversion actually happens.

Reverting to the creation of all matter, the Big-Bang, we find that, at the initial instant only pure energy existed. This instantaneous burst of pure energy expanded in the form of electromagnetic-gravity waves - there was nothing else in existence. Due to random perturbations, irregularities in these waves developed into swirls which created centres of gravity which, in turn, drew into them an increasing amount of the surrounding waves until the increasing pressure caused a rise in temperature. The ever increasing temperature disturbed the now self-centred waves causing a break-up. If a wave becomes broken into individual wavelengths it cannot sustain itself with open ends, it must form a closed sphere - it has become, what we call, a particle but I suggest it is better described as an Electromagnetic-Gravity-Globule – an EGG, the Degg. Matter has formed from waves of energy.

Conversely, if an atom, comprising a nucleus surrounded by a cloud of electrons, should be physically disturbed by collision with other 'particles' (Deggs), it may be partially or completely broken up into disconnected particles and/or waves of energy of short duration. Although nuclear physicists would say this was an over simplification of a complex process, it serves to illustrate how energy can generate from matter. Radioactive materials, such as uranium, have so many components to their atoms that they are unstable. Particles are continually being ejected in an attempt to decay to a more stable material. If sufficient radioactive material is combined in one mass the ejected particles will bombard surrounding atoms creating a chain-reaction with the resulting explosion of energy – hence the atomic bomb.

What triggered the Big-Bang?

Cataclysmic explosions are caused by gravity. All universal electromagnetic-gravity waves are being drawn towards each other but, due to the inertia of a previous reaction, are tending to expand apart. This subtle conflict is made more complex by the random perturbations locally deflecting the otherwise universal expansion to create 'hot-spots' where the accumulated gravity exceeds the rate of general expansion. As the waves of energy come closer to each other the temperature starts to rise because temperature is, simply, a measure of sub-atomic agitation. A cross-section through space would show greater agitation as the waves come closer together. The temperature increases to a level where 'particles' - broken waves formed into Deggs (Duffill's Electromagnetic Gravity Globules) - are created and then destroyed, re-releasing energy. This is the essence of nuclear burning which creates a back pressure balancing off the accumulated gravity. Eventually, all the material is converted into energy which then bursts out thus forming the expansion of pure energy from which it all started.

How did it all begin?

If we examine the graphical representation of a pure sinusoidal wave - any wave, electromagnetic, sound, water, whatever - we see that, at some point the wave has no value. From there the magnitude increases (voltage rises, pressure grows and height swells) to a maximum and then decreases back to zero and beyond to a negative value equal, though opposite, to the previous maximum. The wave then increases back to zero and beyond, to repeat the cycle ad-infinitum. Similarly, the expansion of EMG waves from the Big-Bang increases in radius at a decreasing rate, due to gravity, until the balance is reached where gravity overcomes the explosive inertia; from thereon the boundary of this 'ball' of waves decreases in radius at an ever increasing rate until it is reduced to zero. That half-wave is the measure of the lifespan of the universe because; at this point (literally) all matter has been compressed back into pure energy. The intense heat generated will cause another Big-Bang triggering off another expanding universe. *This new universe should logically develop as the inverse of the current one - with the dominance of antiparticles rather than particles - in order to complete a new grand wave cycle. However, that is conjecture.* The answer to the question is that the cycle of universal expansion and contraction is continuous; it has neither an end nor any beginning. We should, therefore, talk of re-creation rather than creation.

The foregoing theoretical sequence has been propounded – and rejected – by scientists in the past: however, it is here suggested that the gravitational component of electromagnetic waves not yet formed into matter (presumably previously discounted) *is* sufficient to cause this cyclic sequence.

With these answers to the birth and death of the universe and to the many philosophical questions perceived earlier, the sense of the hypothesis propounded here is illustrated.

Chapter 15

Thinking of the Effects

Before examining how the concept of the fundamentalism of waves truly forms the basis of everything and every action, let us first of all take into account two aspects.

When the principle characteristic of a person is referred to, it is not implied that there is only one wave-form guiding the brain's function; there are many linked, sub-characteristics building to the whole. Diagram 11, below, indicates how many waveforms, from multiple sources, can be combined to form a complex compound characteristic.

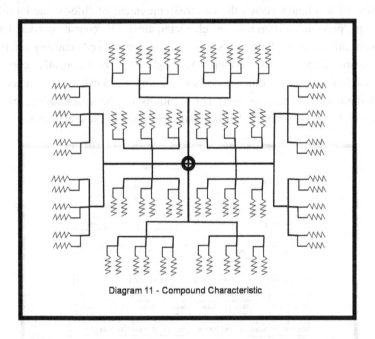

Diagram 11 - Compound Characteristic

As previously mentioned, the relationship between all the sub-characteristics is similar to the relationship of individual files in a cabinet. Consider a document to be a sub-characteristic related to a particular aspect of human activity: although it forms part of the composite structure of the principal characteristic, it need not necessarily be in direct harmony with every other sub-characteristic. After all, the files in the top drawer of a cabinet need not be closely related, or directly compatible with, those in the bottom drawer. They do, however, have a relationship – they are all *your* files.

Each of these sub-characteristics reflects one's attitude to particular spheres of experience. For instance, we have differing views of our geographical environment; some enjoy the aura of the city while others enjoy the openness of country living; some display physical leanings by engagement in sport while others prefer the more sedentary pursuits of academic study or the simple pass-time of TV viewing. Other sub-characteristics will result from the tendency towards managerial duties versus that of manual labour or the pursuit of a trade. The discipline of a religion versus the carefree enjoyment of 'life-on-the-loose' is a prominent potential difference in characteristics. Of course gender has an important influence on how life's experiences are viewed; causing alternative characteristic elements to develop. Although race is frequently considered irrelevant, it does greatly influence one's viewpoint due to cultural differences.

Just think of the multitudinous combinations of the characteristic elements mentioned above: Diagram 12, below relates just some of these elements.

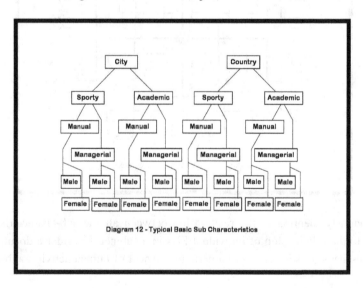

Diagram 12 - Typical Basic Sub Characteristics

On the one hand we have a city orientated, sporty, male, manual worker; on the other we have a country loving, academic, female manager. These variations of general characteristic would be further spread if we introduced other factors such as marital status, physical stature, age, skin colour, hair colour, and so on. The variations are endless.

All these, and more, sub-characteristic elements must be in general harmonious relationship for the formation of a stable character overall. Where there is conflict between the elements in close relationship, instability in that sector will be the inevitable outcome. This 'sectorisation' of characteristics does surely lead to the complexities of personality demonstrated by certain individuals; less than harmonious relationships between sectors are probably the explanation of mood swings or, in extreme cases, split personalities such as bipolar disorder.

It should be noted that throughout the preceding chapters, the word "characteristic" is used in preference to "personality". Whereas the collective characteristics form the personality, it is important to accept that the characteristics are actual electromagnetic waveforms resulting from the person's lifetime of experience – not just an *apparent* display of attributes.

'Sectorisation' of the brain so that waves related to certain particular spheres of experience can be compared for acceptance or rejection, account for the development of sensations. Take for example leisure; many people enjoy the works of Shakespeare; to read or listen to his plays brings pleasure. This no doubt stems from their educational upbringing when the initial seeds of understanding of literary merit were introduced - forming a sub-characteristic in the appropriate sector. Repeated experience of such works strengthens that sub-characteristic to the point where joy or satisfaction is felt.

Those who have not been so endowed cannot find any pleasure in the Shakespearian experience – often considering it to be gobble-de-gook, simply because they have no sub-characteristic to which it can be fitted.

Particular sports generate great excitement to some while others find no enthusiasm in that contest but are ardent fans of some other pursuit; football v cricket, motor racing v gliding, etc. The more one identifies with a particular activity, becoming more aware of the rules and of the combatants, the stronger becomes the sub-characteristic associated with it – leading to harmony with those of like mind. It can well be that you share a compatible association with a particular person, or group of persons, regarding one subject matter but have conflict with the same person on another; sport v politics, religion v science, etc.

Humour is the classic example of the acceptance or rejection of new input. Types of humour vary wildly, be it accidental or contrived. Scan the television channels for programmes which suit your own developed sense of humour and I suspect you will reject more than you accept. One's taste also tends to vary with age; our humour sub-characteristic becomes more sophisticated. The slap-stick comedy enjoyed as a child becomes over-written by more subtle, suggestive humour. However, if the received input fits your developed profile and has sufficient strength of character, the resultant development of your humour characteristic will, no doubt, cause a physical reaction – laughter.

Similarly sectors devoted to fear and anxiety, satisfaction and pride, jealousy and hate, general knowledge, specialist knowledge, and so on, all develop sub-characteristics which are drawn upon in conjunction with each other. It must be remembered that, due to electro-magnetic induction, each sector will have some influence on adjacent sectors. This will lead to a general tempering of mental reaction of a whole zone of the brain. It is unlikely that a person of steady, thoughtful disposition would react violently to new input. Similarly, someone of an impetuous nature will not give adequate thought to what lies behind new input before taking action.

The second aspect of characteristic development of which we need to be aware is the learning curve; it takes time to assimilate new knowledge and it takes time to substitute a revised appraisal of experience. In the engineering sphere of electromagnetism, the technical name given to this learning curve is 'hysteresis'. A hysteresis curve portrays how absorption of new conditions increases as the time that the signal is applied increases. After a while, and depending on prevailing conditions, saturation occurs - absorption has reached its maximum and nothing further can be learned under the existing circumstances. If then the input is withdrawn, the level of learning will commence to decay, very slowly at first but then with gathering momentum. The level of reduction will not normally reach zero, there will be a residual element remaining - a basis for re-learning. If, however, a coercive force is applied, that residual element will be unlearned and an opposing element learned in its place. If this new experience is in total opposition to the previous, then the learning curve will be similar, but it will be a mirror image of the original in all aspects.

The point is that it requires a definite opposing experience to negate the one already established – as is the case with homeopathy. This draws our attention to the dubious practice of brain-washing, but it also underlines the importance of what is learned early in life - before the characteristic has attained a high degree of form. Our childhood learning becomes a basis on which to build a developing characteristic unless new, totally opposite experience cancels out that which has already been established.

> In the epilogue of his book *The Sun, The Genome, and The Internet* Freeman J. Dyson recalls the day in 1977 that a computer programme, *Deep Blue*, was able to defeat the great chess champion Gary Kasparov. The author draws attention to the fact that the software had learned from Kasparov's many games over the years and had an understanding of Kasparov's characteristics. On the other hand, Kasparov had only the half dozen games of the short contest to learn anything of the characteristics of *Deep Blue*.
>
> This is a clear example of how the act of learning by repeated experience in a particular field builds a characteristic.

With these points in mind, we will now look at the consequential effects of the development of characteristic in the human mind and, to a lesser degree, in the characteristics of non-human bodies – even in the solid matter of our environment.

The development of an individual's characteristic will, to some degree, have an effect on others; it all depends on communication and strength of character. In a family, the general policy and outlook stems from the leader of the household, the father or the mother – whoever is the stronger in character rather than who ever is stronger physically. A community's characteristic will change in accordance with the leadership and example set by its 'elders'. Going further, a whole culture will progressively be modified due to the accumulation of knowledge and experience.

For example, today's cultural climate is deeply entrenched in the conservation of the environment. This contemporary culture is itself a developed characteristic which has evolved due to satisfactory layering of harmonious vibes stemming from 'global warming', 'environmental pollution', 'high speed worldwide communication', and other – comparable facets of our modern era; enabling development in accordance with Darwin's principles of evolution.

It is the developed culture of the military which enables man to kill man; a soldier is indoctrinated (they would rather say trained) into the need to shoot the 'enemy' in order to defend one's way of life – one's culture. Soldiers do not normally simply kill the enemy, they immobilise him to defend what is believed to be right. This is why it is right that those soldiers that *do* kill – solely for perverted pleasure should be charged with murder; they have been operating outside the developed culture.

In Hitler's Germany a culture of murder was established under the umbrella of ethnic cleansing, this, however, was reversed after that country's defeat in 1945. With such exceptions, cultures of the world tend to progress in one direction. This can be readily understood in the context of this hypothesis. The addition of harmonious vibes (waveforms) is accepted by the existing characteristic causing a modification to a more sophisticated aggregate. So it is that our cultural development is based on previous experience which has proved satisfactory – any unsatisfactory excursions tend to be automatically rejected.

We tend to forget that the current cultural norms were not always so; we have changed our attitude to slavery, to racialism, to religious tolerance, to colour and class. Although in some areas of the world there is still plenty of room for improvement, the general trend – accelerated by modern communications and by increased international personal contact, the trend is favourable to universal acceptance.

There is, of course, one notable exception – that of the conflict between Islam and the non Muslim world. Here we have another example of the progressive development of a characteristic – the cultural characteristic associated with religion. Hard-line adherents to fundamental religions such as Islam, or Orthodox-Jews, and many Christian sects, have been indoctrinated over generations to accept that legendary scriptures should be taken literally. To them, the dogma is 'right' and anything not following the rules is 'wrong'. This, of course, is dangerous because it does not allow for the natural progression of the culture due to generations of experience. The imposed dogma does not allow for any tolerance; consideration of alternative views is not allowed. The long standing disputes in Northern Ireland between Catholic and Protestant, and the post WW2 disputes in the middle-east between Palestinians and the Israeli, are classic examples of this intolerance stemming from adverse cultural development.

If these factions could be made to realise that all of us on this planet - and any other body in the universe, whether living or inanimate, stem from the same source and are linked by the same radiation of energy waves from the creation which is the All-Enveloping-Influence we call God, then perhaps – just perhaps, by tracking back through the evolutionary chain, all factions would begin to accept we are all brothers and sisters.

The relevance of the All-Enveloping-Influence of God's waves has frequently been touched upon in the preceding pages in relation to the human characteristic; but what of the characteristic of inanimate materials – the bricks and mortar of our physical world? In the context of this hypothesis we can, for instance, understand how places become holy; why ancient cathedrals, temples and other religious buildings or special locations exude an aura revered by visitors.

Over the years, adherents come to these places of worship in search of spiritual satisfaction - they wish to tune-in to God's vibrations. To do this they collectively pray to a common theme as well as invoke the harmony of music by singing or chanting in unison; all present become of like mind. This, in turn, triggers a resonance between worshippers, developing a powerful signal on the same waveband as the energy from Creation. As with two lovers; the worshippers are temporarily adjusting their personal characteristics to a united spirit. This is the basis of the expression of having a love of God.

The aura of the combined thought-waves of the congregation is of sufficient magnitude to enter the very fabric of the building. The stone walls, the wood-work, the roof and the floor – being basically formed from waves - become infected. Like the homeopathic water, the waves of emotion change forever the characteristic of the physical matter of the structure. The building itself now has a unified characteristic; one which closely fits the ethereal emanations from The Creation, since to tune in to those waves was the very purpose of the congregation. The **building** now has a soul. Resonance will now take place: the ethereal characteristic bouncing back and forth between the mind of the people and the body of the building. Over time, the residual effect gains strength to the point where similar waves are re-radiated back into the enclosed space even without current positive fresh input. Consequently, when anyone visits the interior where the aura is focused, the presence can be detected - particularly by those who are basically in tune with God's All-Enveloping-Influence; this enclosed space has become holy.

A similar effect to that of holiness is homeliness, or friendliness, which can be generated in an ordinary home, a place of work, or a place of relaxation and rest such as a guest-house or a hotel. This is achieved by the exchange of love and/or consideration for others being practiced within, creating harmonious vibes which become absorbed in the surrounding fabric of the building such that, as soon as one enters a favourable aura is detected. Equally, but to opposite effect, a stressful, hateful and generally uncomfortable environment can be generated by the inconsiderate or otherwise alien behaviour of the occupants. To enter a police interview room or detention cell will create a feeling of foreboding even if you are not personally the subject of enquiry. A dentist's surgery would be most unusual if it could be described as having an appealing aura, even though it might be the place of pain removal.

One of the main reasons why shopping malls have become so successful, and therefore more common, is that a pleasant aura is deliberately created by the provision of a clean, warm, adequately illuminated environment backed up by suitable music to create a relaxed atmosphere. In such conditions, shoppers enjoy the experience of seeking out their purchases; they are in a happy mood even though their cash is being solicited. Others around them are similarly having their materialistic desires satisfied. Subconsciously all are joining together in common purpose so that collective waves of satisfaction are being radiated. The very fabric of the building absorbs this general transmission and re-radiates it as a pleasing aura; putting shoppers in the right frame of mind to make more purchases.

While on the subject of inanimate objects having an aura, let us consider the moon. As well as creating the tides, there seems little doubt the moon does have an effect on people - some people more than others. How can this be when we are all now aware that there is definitely no man up there; it is a totally inert lump of rock and dust? We do know, however, that that lump of rock has a gravitational effect; hence the tides. Let us now recall that, according to the given hypothesis, the EM waves forming all matter also develop a gravitational component. It is the aggregate of these components which accounts for the total gravity of any body. Not only do electromagnetic-gravity waves form the basis of all matter, they are also the signal link in the brain/mind/soul complex. It is, therefore, quite understandable that the bias created by the presence, or otherwise, of this great gravitational body will have the potential to influence the brain/mind-waves. An unstable mind might suffer distortion causing possible irrational behaviour due to the additional mental pressure generated by the presence of the moon's EMG waves. The romance of the moon is highlighted by two lovers having similar thoughts; their minds are drawn in the same direction by the awareness of this illuminated electromagnetic-gravitational ball.

Considering how our nearest planetary brother can affect the workings of susceptible minds suggests the possibility that, when at our most vulnerable – the moment of birth, the disposition of the solar planets may indeed influence the initial establishment of the human characteristic due to the combined EMG effect. That observation perhaps gives encouragement (rightly or wrongly) to the adherents of astrology – although I cannot imagine how anyone can correctly attribute the forces of the various planetary bodies to any particular human characteristic.

Unlike bricks and mortar or other inorganic matter, plants and animals are able to physically respond to the presence of an aura. Responses of plants are somewhat limited but they do react to the electromagnetic radiation of the sun. Some plant lovers also insist that responses to words of tender loving care have been observed, although this presumably is not the actual words but the transmitted feelings of the guardian.

Because of their more comprehensive strings of DNA, animals have a much greater ability to respond to stimuli than plants. We are aware that horses, cattle, household pets and others can sense danger and have been known to alert humans to some unseen hazard. On the other hand, although they are unable to directly communicate with us (or we with them) pets do sense the aura of our affection; we do bounce harmonious waves between their minds and ours. Infant animals, deprived of their mothers, build different characteristics from those of their ancestors if brought up by well meaning and sympathetic humans; they absorb certain elements of our own characteristic which would not normally be compatible with an animal reared entirely in the wild. It is for this reason that professionals in the field of animal husbandry seek to minimize human contact and return animals to a natural environment as soon as possible – before they absorb any elements of the human characteristic..

Conversely, trained guard dogs and undomesticated animals do transmit waves of conflict towards humans and other non family animals. These are usually interpreted as fear and aggression. We as human beings, being rather more conscious of induced waves, whether harmonic or conflictic, tend not to be so extreme; generally we live and let live. This, of course depends on the establishment of a stable, satisfactory characteristic. We are all aware how nasty children can be to each other; this is no doubt, due to the fact that they have not yet developed an 'adult' characteristic – they do not have adequate consideration for others.

For years behaviourists have argued as to whether or not a youth's bad behaviour can be blamed on the lack of loving parents. This hypothesis clearly indicates that, for a stable character to be developed, consistent input of harmonics is absolutely necessary in the early years of the formation of a person's characteristic. The caring vibes of **considerate**, loving parents will, undoubtedly, build a wave-form with a sound characteristic. Consideration is a very important element to develop in order to avoid brain-washing children, either by accident or by deliberate act. If the parents are, for example, fanatical with regard to their religious activities, they may quite genuinely believe that their outlook is the **only** way of viewing life. This is the outcome of a characteristic having such a narrow spectrum that all other views will be rejected as 'conflictics'. If one does not accumulate a variety of harmonics – a breadth of exposure, there is little probability that a new experience will be able to be compared for a suitable fit into the general characteristic. A child brought up under these narrow circumstances will lack judgement. Therefore, when faced with an unfamiliar, possibly critical experience in life, he/she then will suffer serious confusion due to having no comparative element within his/her characteristic. This confusion could distort the characteristic in a manner which society would then class as a mental disorder.

Being brought up with a limited outlook can have very serious consequences, not simply at the personal level but at international level. As already highlighted, the whole world is aware of the political difficulties being experienced in the Middle East; the basis being the intransigence of the Israelis and the Palestinians towards each other. The root difficulty in this case is that the nationals of each side of the dispute have developed characteristics which are too narrow; they have never had a sympathetic experience of the other culture. They see their way as being *right* and it has to be right because the opposing view is a conflictic – conflictics are wrong. Testing incoming information to determine whether it is in harmony or in conflict is how we determine right from wrong. Children of these two nations are brought up in totally different cultures, which are largely based on totally different religions, without any effort being made to introduce the other's belief as an alternative; the alienation is, therefore, inbred. The vibes of the Muslims are different to those of the Jews. Whereas the difference is understandable and an objective commentator must declare that neither is right nor wrong, what is lacking is **consideration for others**.

The solution that this hypothesis suggests is that a branch of one's characteristic should be allowed to develop in sympathy with others. If this was done soon enough in the development, a fit would be found for the opposing view. This being only one branch of the general characteristic, it would not mean domination but would allow sympathetic consideration. This sympathetic branch can be developed by encouraging children of both camps to play together, to attend joint functions, even to share classes together when studying non religious subjects such as mathematics, music, physics, chemistry, and even a common language. Only when these generations supersede the present generation of warring factions will genuine peace prevail.

Some years ago in Northern Ireland – a country experiencing chronic conflict – a particular Belfast street was populated by a mix of Catholic and Protestant families. The children enjoyed playing together without regard for their religious background; after all, there is no physical difference between adherents of the two religions and the rules of soccer or netball are the same for both sides. As in any other community, friendships developed, regardless of background. Hard line political groups, however, saw this free association as a wedge opening a crack in their campaigns and brought it all to a halt. They separated the families by threat so that the children would be brought up on opposite sides of a barrier – literally.

Thus the dominant sectarians had pruned off the sympathetic branches of the developing characteristics.

Usually, youthful bad behaviour takes a less serious form. When a person with a less than solid character is released from the strict discipline of a restricted upbringing, he/she is vulnerable to the reception of vibes from mischievous minds of new acquaintances. If the narrow characteristic does not reject these unfamiliar signals, the likelihood is that they will be allowed to distort the existing wave-form; the person experiments with this new experience and finds acceptance amongst his peers. This element is now reinforced and is in danger of becoming a permanent feature unless countered by an appropriate stabilising injection of vibes.

A prisoner being released from jail is faced with the same situation; his recent environment has established norms which do not fit with the outside world. As a result, he or she is particularly vulnerable to rejecting society's accepted cultural norms in favour of those of the criminal world.

A strong but narrow upbringing can be the cause of a distorted characteristic in the same way that the lack of proper parentage causes a weak start to life. Even the bricks and mortar of the environment - the prison, the mental hospital, or even the school, have an effect. As discussed earlier, inanimate materials do also absorb, and re-radiate, vibes - they have an aura. If the aura of one's environment is of a restricted nature, there is no breadth of foundation from which to draw support. With these points in mind, we can see a fundamental basis for the development of a good stable character or an unstable, mischievous character - even one of a criminal nature. It all comes down to waves.

Strict protective upbringing not only has a potential to adversely affect behaviour, but it can affect health. When living in the suburbs of the English city - York, our children often played with the boys across the road. Collectively they would get up to mischief, frequently arriving home filthy dirty. Another boy across the road, living next door to our children's playmates, was never allowed to join in for fear of him becoming dirty which, of course, would be unhygienic. As do all children, ours and their playmates suffered the usual childhood ailments such as measles. These they shrugged off without special difficulty; their immune systems functioning satisfactorily. The child next door, however, almost succumbed to these infant illnesses owing to being too clean. It is suggested that the four mischievous kids had progressively experienced small samples of bad hygiene through playing in dirty water or around filthy drains. Thus they had conditioned themselves so that the body knew what to reject. In contrast, the 'clean' boy's body had no previous experience to know how to attempt to reject the germs to which he was finally exposed; the impact confused the defence mechanism, throwing it off its desirable course. This was an example of a sub-characteristic (the germ defence system) knowing, or not knowing, when to reject a conflictic.

The same principle applies on a more general scale when a traveller visits an unfamiliar country whose standards of health-care differ from those of the home base. The inhabitants of the visited country have become immune to the hazards of the environment – as in Leningrad after the war; their physiology has a characteristic modified by toxins common to their location. This is a similar principle to homeopathic cures. It is for this reason that vaccinations were introduced; to prepare the body for the expected conditions by widening the appropriate characteristic through sampling small doses of toxins.

Single parents are handicapped when attempting to provide the wide based firm characteristic needed at the start of a child's upbringing. Even though the mother may love her child dearly and commendably seek to give all the support she possibly can, her offspring will lack the element of love between two mentors; he/she will not know what the harmony of love is - it is not being observed in the highly active learning period immediately following birth. The stability offered by a pair of loving parents can't be overstated; it straight away establishes a desirable wave-form within the developing brain of the child even before the mind has evolved. The mind then has this firm fundamental to build on.

An additional, inevitable problem is the need of the single parent - whether mother or father - to earn a living. This results in the child being left with minders for significant periods of the active day; this then deprives the developing mind of the continuous subconscious input from the parents' minds. Instead, the child is receiving vibes from new sources which, although to some degree will cause a widening of the characteristic; like the spreading of the estuary of a river, it causes shallowness - the principle element has not yet developed on which to build. Another danger is that if the minder is one of the grandparents, then the input will be more consistent but from a different source. The child will be generating a characteristic possibly with two main wave-forms; that of the parent and that of the grandparent. This duality can cause confusion to a degree dependent on the closeness of the parent to the grandparent. If they are significantly different, the child could have the element of conflict introduced which gives a most unstable basic characteristic even though parents and grand-parents were devoted to the child's welfare.

In cases where life started with two parents who subsequently broke up, an even more unsatisfactory environment is created for childhood development because, not only do the mentoring pair lack the stability of a loving relationship, they have introduced the element of conflict. Due to the child's mind not having yet developed, there isn't a strong characteristic to reject this 'conflictic'. Worse still, this is an ongoing situation which continually reinforces the conflict element in the developing mind. The child involuntarily accepts this element as a norm. He/she will grow up subconsciously sceptical with regard to the stability and harmony that love can bring. This scepticism will, unconsciously, be transmitted to their future children unless a strong love partnership should occur.

> I know of a family who have had a typical weak start to the development of their respective minds because of the unsatisfactory marital relations of their parents. As a result, the children never experienced the harmonious environment of loving parentage. As is the way of life, *like* attracts *like* caused by the similarity of characteristic wave forms. This leads to the siblings of the family meeting siblings of a similar family. One such young lady – by now a single parent - having already suffered a failed marriage, attended a function where she was introduced to a number of people from a different background. This was not a social class difference but a difference resulting from a harmonious upbringing. The new-found associates largely had long standing, loving relationships - an extended family with many marriages lasting fifty years and more. The vibes that the unfortunate young lady received were a revelation to her; she was overwhelmed by the harmony displayed between

members of the extended family - something her mind was unaware could exist. Not just her eyes, but her mind was opened to new possibilities. This is a true account of a young lady's realisation which was witnessed by your author; identities have been deliberately omitted to avoid embarrassment.

The culture in which a person is brought up creates a fundamental base, an aura, which guides the development of one's characteristic. It is obvious that a Chinese person will think like a Chinese not like an African - at least until exposure to other cultures has occurred. It is a misconception that it is only differences in language that create the difficulties of understanding found between foreign cultures. The whole way of life, the traditional behaviour of one's parents, induces certain elements into one's characteristic. Responses to given circumstances differ notably and are dependent on cultural and religious foundations; if a cow is slaughtered in the Argentinean cattle ranges, this will cause salivation at the thought of the impending good meal. If, however, the same thing happens on the streets of Mumbai (Bombay), a great disturbance - approaching horror - will be caused at the thought of the wanton elimination of a sacred being.

A rather morbid example of different cultural responses to similar situations occurred during my residency in the far eastern nation of Malaysia. My Assistant Manager was an Indian, born into a family practicing Hinduism, but was married to a Chinese lady. Sadly, their much loved, infant son died as a result of a tragic accident. As his employers and friends, my wife and I were invited to participate in the funeral arrangements, but this resulted in a very harrowing experience. Due to their different cultural backgrounds, the unfortunate parents had long since found common ground in the Christian faith through the Catholic Church; however, the requirements of the Hindu faith were met by bringing the small body to the family home where he was laid, exposed, on a central table. Much chanting and soliloquising took place

143

which we did not directly understand, but the highly charged atmosphere conveyed to us the common feeling of grief and despair. This was a clear case of perception via mental vibrations rather than linguistic translation. The mental pressure increased when a small, empty coffin was brought in and the child's body transferred to it in front of all those gathered around. After favourite toys were added and flies removed, the lid was fastened down for the journey to the next phase.

The Christian faith was then professed at a service in a Catholic Church; prolonging the emotional stress experienced by all present. The third element, presumably a Chinese custom, was the actual burial where the grieving parents, on their knees, personally scooped the mounds of soil with their bare hands until the grave was full. Afterwards a marker was added followed by the addition of a candle from each mourner. *To add drama, the whole burial took place in a tropical downpour.* We drove the difficult hill-road home absolutely drained; all our energy had been expended in the unconscious act of transmitting so much emotion in concert with everyone else.

There are a number of points we can observe from this tragic tale. First, and foremost, it is a good example of how different cultures respond to a particular situation – they have different cultural wave-forms. It was a good example of resonance; each person's feelings were being transmitted to those around, causing an ever increasing level of emotion within the group – a powerful aura was generated. As already mentioned, mental vibrations transcended any lack of understanding of language – the mind was working on wave pictures rather than word pictures. The exhaustion felt at the conclusion of this tragic day was clear evidence that much energy had been expended in the creation of the waves of emotion – the waves were transmitting actual energy.

Even at the personal level, we do not necessarily respond in the same physical manner when we experience similar emotions. This fact was underlined during an embarrassing incident while managing a hotel in what, to me, was the foreign country of Malaysia. I had engaged a Chinese chef who carried ample written qualifications from a Swiss culinary college. After he had been working for some while, it became apparent that his knowledge of hygiene was inadequate, particularly in the use of refrigerators and freezers. Fearful of having an outbreak of food poisoning, I summoned him to the office for an appropriate admonition. In the presence of my Indian assistant, I pointed out the error of his ways and demanded an improvement. Instead of hanging his head in shame or flushing with embarrassment, he smiled. This annoyed me and I responded strongly - he smiled all the more. I rose from my chair and stormed out of the office, commenting that this was a serious matter not to be taken frivolously. My Indian assistant chased after me to point out that I was making a grave mistake; "the Chinese tend to smile as a response to embarrassment or shame", he pointed out. It was *I* that did not understand, not he. It was also an example of an inbuilt attitude that one's own culture is considered superior to that of others – *a common failing of the North American culture.*

The 16th century French essayist Michel de Montaigne discovered for himself how parochial the minds of people can be if they have not had the experience of travel. Although his cultural excursions were limited to Europe, he found that Germans had different eating and heating traditions to his fellow Frenchmen and that the Swiss and Italians displayed eating and sleeping habits unfamiliar to him and his retinue. They lived in homes which had different architectural features. In essence they were foreign to him.

These differences may seem trivial but, having lived one's life in a restricted cultural environment; a barrier is created; besides the language difference, foreigners *think* differently. Historically, ignorance born out of such differences has led to the subjugation and even massacre of whole indigenous races by invading pioneers; they have the mistaken belief that their way of life is superior. It appears that their basic cultural characteristic is not harmonious to the newly discovered characteristic of the native tribes, sometimes to the extent that a violent reaction takes place. A good example of this is the historic invasion of the Aztec region of Central America by the Spanish; who then set about wholesale slaughter of the indigenous tribes.

Countries such as the USA or Australia have many cultural divergences; not only is there a variety of indigenous tribes, referred to as Indians in the USA or Aborigines in Australia, but the non-indigenous population is made up of citizens with many diverse origins. A variety of European stock has been supplemented in the USA by migrants from the Spanish speaking Central American states, while in Australia the Orient, South East Asia, and the Indian sub-continent provided the migrant mix. Both nations have a selection from the rest of the world. Because the cultural characteristic wave-form differs, a greater understanding, and sympathy, is experienced within each group rather than the nation as a whole. This has the unfortunate effect of causing gravitation: Indians tend to marry Indians, Greeks send their children to Greek schools, Italians gather together to consume pasta. The collective result is that full integration is not achieved.

A campaign of cultural integration has recently been launched in Australia; entitled "Fair Go Australia", it seeks to break down some of the barriers of ignorance by arranging joint functions so that opportunities are presented to understand a little more of our fellow 'man'. It is interesting to note that the opening event was predominantly a presentation of music from some dozen or so different cultures found in the Greater Sydney area; this was a classic example of the harmony generated by music.

A more general example of having biased cultural views based on one's 'localised' upbringing is that of hemisphere. Most of the world's population is in the northern hemisphere and the majority tend to make the mistake of forgetting that those in the southern hemisphere experience inverse seasons. It is also not generally acknowledged that those living in the equatorial regions do not have summers and winters; merely 'wet' and 'dry' seasons. It follows, therefore, that it is unhelpful to refer to worldly events as happening in the summer or the winter; it demonstrates a lack of understanding of the circumstances of others.

What have these examples of interaction with 'foreigners' got to do with waves? They show the existence of a difference in fundamental characteristic wave-forms which, without the appropriate additional element, do not fit together properly.

Branches of one's characteristic should have elements adaptable to the characteristics found in other cultures. The cliché "travel broadens the mind" is based on the fact that we all need to understand others, not just their point of view but the underlying fundamentals of their background - their culture, religion, and the environment of their upbringing. These are all major elements in the formation of their characteristic - the make-up of their soul. Only with this understanding will new harmonics find a match and unacceptable 'conflictics' be deflected in favour of a more harmonious and constructive association.

In the opposite sense, the acceptance of other cultural trends can lead to dominance and lack of the colour of individuality. The contemporary world is well aware of the dominating influence of the American way of life that has largely been brought about first by the Hollywood films and currently by the spread of television and the Internet. Finding the media of film and television to be entertaining and potentially informative, we do not automatically reject the stereotypical elements contained therein. Modern generations are, from a very early age, bombarded with images not necessarily appropriate to their way of life; as a result it is now a lot more difficult for a person's character to be established on a firm cultural foundation.

Slowly and inexorably the stories told on the screen are introducing nuances peculiar to the way of life of the U.S.A. Even their adopted language differs from the English of the mother country; the grammar differs and the spelling differs. Filmgoers and television viewers in non-English speaking countries generally learn the language through the media. It is, therefore, American English which is learned. More importantly, while learning the language, they learn the culture of America. If a person accepts that culture, he/she is likely to follow the way of life portrayed. The logical outcome is those that follow are being led; in this case - being led by America. This may well have started out as an accidental consequence of the film industry's dominance by Hollywood but, I suspect, it is now political policy to promote this trend.

It might be thought to be desirable that the whole world should think alike but this approach does not necessarily bring harmony. Many disputes arise within a family if the family characteristic is not built on a sound foundation. The progressive build-up of one's characteristic wave-form towards a firm harmonious aggregate is quite subtle: it does not normally allow unwelcome elements to

147

inveigle their way into the mix; it requires positive thought to change the direction of development of a sound character. So the 'thinkers' will always differ from the 'absorbers'. Regrettably, in this complex age in which we find ourselves, masses of people spend a large proportion of their day thoughtlessly viewing; allowing their minds to be progressively conditioned – to have their characteristic wave-forms adjusted by the frequent input of the culture portrayed on the television.

Advertisers take full advantage of this situation by conditioning the viewer to accept materialism as the normal aura of a mature person; developing the 'must-have' syndrome. This leads to the inflationary spiral experienced by the modern world as wage demands attempt to keep pace with the perceived needs of the populace. Perhaps the introduction of mass production, which has led to the world of disposable products, was the trigger to the change in outlook of the, so called, western world. My personal recollection of pre-WW2 years is that inflation was virtually unheard of. I was only a child at the time but I could always buy a small bag of sweets for ½ penny and a loaf of bread was always 4½ pence.

An incidental feature of "Americanism" is its disregard for the international standards of measurement (Système-Internationale-d'Unités – referred to as SI) which was established a number of decades ago. In general, Americans still use Fahrenheit in preference to Centigrade, Gallons instead of Litres, Feet in place of Metres, and Miles in lieu of Kilometres. If one attempts to persuade a person familiar with Fahrenheit to use Centigrade – perhaps, for example, when evaluating the climate – the common response is "Oh, I can't think in Centigrade". In the context of the foregoing explanation of how our personal knowledge is accumulated, we can understand the truth of that statement; their characteristic wave-forms do not include Centigrade and so a matching fit cannot be found. The same applies, of course, to Litres, Metres and Kilometres; they do not exist in the mind of many people; they are simply words which they haven't bothered to evaluate. In these cases the mental response is similar to when a monolingual person is confronted with foreign languages; the brain has to adopt the laborious practice of translation. Again, we have evidence that what was first learned dominates subsequent experience until repeated opposing introductions progressively re-educate the brain/mind duality by overcoming residual knowledge.

The greatest concern created by the spread of "Americanism", however, is the development of the belief in the characteristic that their way is *right*. We discussed earlier that the conception of what is right and what is wrong depends on how the matter under consideration fits with one's established characteristic. We each see a situation from a view-point dependent on our experience of life so far; this tends to disregard alternative cultural upbringing if the person has no feel for other life styles. Consequently the inconsiderate mind is bound to feel that his culture is superior and should be adopted by others. In the USA, some years ago, McCarthyism developed as an antidote to the feared Communism; currently we appear to have an anti-Islam campaign.

The imposition of a particular characteristic on another is arrogance. An arrogant attitude does not lend itself to consideration of the opinions of others - an essential requirement for harmonious relationships.

A particular form of undesirable cultural development is pornography. It is quite natural that the sight of a shapely human body of the opposite gender should excite the mind of a sexually active person; this is nature's way of initiating procreation. It is also understandable that in the mind of an immature person the experience might trigger pleasurable bodily responses which he/she is unable to control. Difficulties develop, however, if the person concerned seeks more and more of these pleasures; this element of his/her characteristic is being strengthened by repeated experience to the point where titillation has become the norm. As with the abuse of drugs, this normalisation can become boring; resulting in the desire to upgrade the level of stimulation by the witnessing of even more sensationally lewd acts. By doing this, the relevant sub-characteristic becomes progressively stronger until it dominates the mind of the subject person; who has now become mentally deranged.

In these times, we have the benefits of technological developments such as the Internet, the Camcorder and, recently, the mobile telephone which can take and transmit moving colour pictures. Regrettably, these have become the tools of the widening culture of pornography. In the past it was a very risky business to photograph sexual events because it was difficult to process the film without being exposed to authorities. With the advent of digital still and movie cameras, images can be created and copied without the need of specialist services. The Internet facilitates the distribution of the material to an ever widening audience able to view it in secret. The immature are able to ogle these images regularly in the privacy of their bedroom, cultivating this branch of their characteristic. The culture that results has developed a warped appreciation of what was a natural element of human behaviour. Understanding this leads one to realise that such devotees will be unable to consummate a loving relationship in the normal way; they will lack the element of their characteristic based on harmony and seek only sexual gratification. It is more than likely that this situation is largely responsible for the trend away from stable marriage to that of divorce.

Drug abuse is another form of cultural development having undesirable consequences. Apart from the physiological damage caused by the smoking of tobacco, how might the brain be affected? On the one hand, cigarettes seem to provide stimulation yet, on the other they calm frayed nerves. It appears that the drug content, although relatively mild, suppresses the brain-waves; almost as if a cloud develops between brain and mind. The mind probably does not respond so readily to the brain's activity peaks. This reduction of response is like oil on water; the waves are smoothed out, bringing calm to a disturbed mind. So long as these effects are of minimal severity, there is probably little harm done, if any; one can perhaps visualise circumstances where the calming effect is actually beneficial. Much use of this effect is made by specialists using stronger drugs such as cocaine in the dentist's surgery or morphine on the battlefield. Care has to be taken in these cases to prevent an overdose leading to permanent damage; if the waves are smoothed too far, all character can be lost.

The apparent stimulation from drug use comes about by the reduction in the strength of one's inhibitions; the mind is free to explore avenues previously held in check by the memory bank of life's experience. Stronger drugs, however, present an extension to the principle; discovering surreal images can be considered pleasurable but there are dangers. It seems to the author that the apparent cloud created between brain and mind increases in intensity with the strength of the drug; the mind progressively loses contact with the controlling influence of the brain. With all restraints removed, the mind is free to wander aimlessly amongst the sensations currently being received. In the ultimate condition, a total barrier has developed between brain and mind – the person has lost his/her mind.

> In her book *The Private Life of the Brain* Susan
> Greenfield, an internationally acclaimed
> neuroscientist, draws attention to the effect of LSD
> and other drugs of abuse [creating] "a state resembling
> childhood, where one is upset or excited by minor,
> meaningless events, and [being] very vulnerable to
> suggestions and to literal images, without the ability to
> buffer experience with reason."
> Chap.4 p88 '

In some ways television viewing can have a similar effect on the brain/mind
relationship as do mild drugs. Absorption of soap dramas; game shows; home
improvement; human challenges; and the pseudo documentary type of magazine
programmes, all combine to flood the brain with a confusion of sensations
irrelevant to the normal life of the viewer. Adding all these to the memory bank
overwrites the characteristics of true experience. It can be compared to sand filling
in the shape and form of a beach, or drifting banks of snow; what lies underneath is
obscured, one cannot build on such a layer without cutting through.

In the opposite sense to the dullness caused by drugs and pointless TV viewing,
fear causes a stimulation of the brain as a reaction to the introduction of a
conflictic. If an unfamiliar experience occurs, the mental reaction is to reject it but,
because it is unfamiliar the mind does not know how to deal with it other than to
respond without proper thought.

Members of the armed forces, together with the police and firemen, are quite
likely to be faced with life threatening situations. They need a high level of training
with particular reference to the what-if scenario which introduces potential threats
so that the person is familiarised and will respond rationally. In effect, a branch of
the characteristic is developed to recognise the various potential conflictics and
find the key to respond.

> I have had a personal example of the benefit of excellent
> training in dealing with emergency situations. As a fighter
> pilot, I suffered simultaneous multiple failures of my aircraft
> when at 33,000 feet but, although I was concerned for my

safety, I did not suffer uncontrolled fear. I applied the
knowledge of the aircraft systems I had acquired and
progressively dealt with the many resulting problems. After
completing a difficult landing, my confidence in my abilities
rose so that I accepted my next assignment without
reservation. My reflective interpretation of the incident is that
my memory bank had expanded to incorporate, firstly all the
training, then the experience of responding to danger and so
my character had grown.

Anxiety differs from fear in that the brain refuses to accept the conflictic but does
not oppose it; in effect it closes down. The new – undesirable, experience of
anxiety only seems to adjust the characteristic when the cause is finally over. In the
meantime, while the mind refuses entry, the person is likely to appear morose or
irrational. This is because the normal characteristic is being restrained. Perhaps you
have noticed that the physical sensation associated with fear is felt in the heart
while anxiety troubles the stomach.

In previous chapters, we have already discussed the harmony of music but it must
be realized that hearing music is more than simply a pleasurable leisure activity.
Since the pressure waves of sound are converted to electrical signals, the wave-
form received is a direct contribution to the activity of the brain and, as a result, the
mind.

It is now well-established that music has an effect on our behaviour; it sets a
mood for the environment. Marching music tends to promote physical activity; fast
music creates excitement while slow music has a calming effect. Music therapists
have learned to use these qualities to counteract some illnesses. Although this has
been mainly through stress-reduction due to the mood setting ability, it has been
found that even the physiology of the body can be directly affected. It appears that,
as with 'touch', the aura is directly receivable by energy centres within the body.
The brain apparently does not directly control all functions; there is a degree of
autonomy apportioned to the control points of which, according to eastern
philosophy there are seven.

Apparently, although very experienced in their field, music therapists do not fully understand why music has the effect that it does. If we consider the matter in the context of the characteristic wave-form, we appear to have the answer. The electromagnetic waves resulting from the conversion from musical pressure waves are available to be accepted or rejected by one's established characteristic. If that mindful wave-form has been distorted in some way, perhaps due to stress of one sort or another, the introduction of a countering harmonic element can reduce the dominance of the deformation - bringing the characteristic back into shape. Prolonged or repeated absorption builds the strength of the therapeutic signal - the characteristic is learning to accept this beneficial element.

A very interesting connection between music and the inner sanctum of the brain is seen in relation to a comatose person. Various cases have been documented where a person in a coma has been subjected to their favourite music, or the repeated voice of a loved one, and benefited from it. It would appear that, although they have been totally unable to respond, their subconscious mind was still receiving the vibes and the brain was taking in the sense of the communication.

Because the harmony of music is beneficial to the general personal wellbeing, it has long been recognized that singing can be a very healthy pursuit and may lead to prolonging life. There are various probable reasons for this: If a person sings on a regular basis, such as being a member of an organized, disciplined choir, continual harmonious input is taking place. Being in a group of like minded people triggers resonance of beneficial vibes with those around – including the audience when performing at a concert. *There is also a physiological benefit of trained singing: truly deep breathing, which is necessary in order to sustain the choral sound, leads to the ample ingestion of energy-giving oxygen which circulates to the whole body.*

The ability of the mind to compare new wave-forms with known profiles leads to an understanding of the aptitude of some people to be fluent in more than one language. As mentioned earlier, the sounds arriving at the ear are converted into electrical signals for processing by the brain. If one has studied another language, either formally or empirically, profiles of the words learned and the syntax of their use are established in memory. It is suggested that these profiles will take the same form as the appropriate words in the parent language. Why? Because we do not actually think in words; we think in the *meaning* of the words. The meaning of

"Oeuvre la Porte" has the same meaning as "Open the door". If the person is fluent in French and English he/she would have no need to consult a mental dictionary to find the equivalent words.

On the other hand, if the person is not fluent he/she has to go through the time-consuming procedure of scanning a mental dictionary looking for each word. This is why, to the learner, foreigners seem to speak far too quickly; they don't give sufficient time for the student's translation.

An extension to this concept of thinking in the meaning of words takes us back to earlier comments regarding mathematics. If one is unfamiliar – not fluent in the language of maths – then much searching has to take place to translate a problem into a manageable form. Conversely, those who are familiar with the subject and its procedures will have the ability to think of the meaning of the data presented and be able to interpret it appropriately. This is all possible because a library of interlocking profiles have been established by training and experience.

We must always remember that these profiles, as with all characteristic wave-forms, decay with time unless re-energized. One's childhood learning of a foreign language proves to be of limited value when put to the test 30 or 40 years later – we become 'rusty' and could never be mistaken for a native. As with everything in life, there is usually an exception; when journeying on a train in Russia, back in the 70's, my wife and I met two fellow travellers with whom we shared a table in the buffet compartment. We couldn't speak any Russian but wished to engage in conversation. It transpired that the gentleman sitting opposite to me had learned some English at school but had not had any opportunity to use it. Rising to the challenge he struggled to explain the situation, going on to state that he was a policeman on leave. It was very notable that the more he tried, the more fluent he became; he had found the mental pathway to a decaying store of his unused training.

> *This exercise in mental recall led to an amusing incident. The gentleman sitting alongside me was unable to speak any English at all and, it turned out, did not know the other Russian passenger. The policeman explained to him what was going on, so the mystery man tried to explain what he did for a living. He showed me a picture from his wallet depicting*

himself in a tuxedo, standing on a stage. "You're an entertainer." I suggested. This was relayed by the policeman. Mystery man then conveyed something about my wife's underwear, which seemed very forward of him until we discovered that he was searching for the word 'silk'.

"He wears silk", we concluded as we suggested; "A lawyer"; "A jockey"; but the answer was always "Niet". I leaned back in the seat making the derogatory remark "He's a bit of a clown, this bloke." "Clown, Clown!" came the excited reply; he was a clown with the Moscow State Circus.

With laughter between complete strangers, we were then treated to share their refreshment – harmony ensued.

These pages seek to reveal that electromagnetic waves form the comprehensive link between all matters, whether physical or philosophical – from the "Big-Bang to Brain and Beyond". Whereas these waves are usually very constructive with regard to physical matter and to mental awareness, they can also be very damaging under certain circumstances.

The liberation of energy during a nuclear explosion is the conversion of matter back into waves of energy. Some of these electromagnetic waves are the very energetic gamma waves, which being of high frequency, have lethal penetrative qualities. This radiation would be mankind's undoing in the event of a nuclear holocaust.

Less lethal, but quite insidious are the electromagnetic radiations from electrical installations in developed countries – particularly in the home. The author does not know if awareness of this danger is being suppressed but has definite experience of its danger. Some while after moving to a new residence, my wife started to become chronically ill. A friend reported that her own similar experience had resulted from the bad location of the main electrical feed cable to her home. We had our home checked by a radiation meter and found that the main bedroom suffered from very high levels of EM radiation. This was traced, quite clearly to the location of the distribution panel which was on the outside wall of that very room. After we had relocated to an alternative bedroom, as far away from the panel as possible, my wife's health progressively improved; she subsequently made a full recovery.

We must remember that, not only do the brain and the mind function via the transmission of electrical signals, but the whole body is comprised of atoms built up from electromagnetic waves. It is no wonder that continual bombardment by radiation from a close source can be the cause of physical malfunctions.

The application of mind over matter is unreasonably treated as a fiction by some. The human body is, without doubt, a highly complex assembly but, in the final analysis, it all comes down to waves – electromagnetic waves. The brain relies on mind waves, under the direction of the soul, to carry out its many body control functions. Although basic functions are autonomous, abnormalities require decisions to be made by the control centre – just as is the case in business management. If the brain receives a pain signal from the leg, for example, it might simply instruct the speech organs to cry "Ouch!" On the other hand it might instruct the injured muscle to take it easy for a while; to walk with a limp until the pain has subsided.

This trivial example gives us the clue that a really strong mind can persuade the body to deal with more critical problems. It is believed that some very strong minds have been applied to overcome such extreme conditions as terminal illnesses. Many can quote cases of sick or ageing people living until an important event occurs. My own mother survived on her death bed until **both** of her sons had travelled to be together at her bedside, whereupon she passed peacefully away.

If one focuses the mind on a problem such as a pain or an illness, and imagines the remedy, the brain will send the necessary chemical antidote to assist with the cure. You might question how one can imagine the remedy when the remedy is not known. One must give credit to the brain to know, at its most basic level, what is going on. If you truly, dearly **wish** to be cured, thinking deeply about the malady, you will be focusing on the remedy – the autonomous level of the brain and the energy centres will receive direction and generate the appropriate responses. Alternatively, imagining a chemical oozing into the affected area and 'washing' away the disease will have the same effect. It requires imagination and **belief** in the abilities of your brain and body to achieve the desired result.

This thesis suggests we should realise that the electromagnetic structure of the body and all its fluid content can be influenced, both positively and negatively, by the electrical activity of the brain. This is particularly so when one remembers the possibility of resonance between the signals to and responses from the control centre.

In Chapter 8 we considered how various sub-characteristics of sectors of the brain fused in various groupings to form a combined characteristic ultimately creating the personality of the individual. This same principal appears to apply at the sub-atomic level. The electron cloud surrounding individual atoms vibrates in a particular manner – that which is characteristic of the particular element. If, however, that element comes into close association with another element, the vibrations will be modified to create a new characteristic peculiar to the combination of the two elements. Hence if we arrange for a couple of hydrogen atoms to merge with an oxygen atom, the result is not a mixture of hydrogen and oxygen but a new material having a totally different characteristic to the components; this we call water. The very toxic elements sodium and chloride are combined to create a new, quite palatable material – common salt.

So it is that, even at the sub-atomic level, basic elements have character. If we go further, we might even interpret these characteristics as crude 'feelings' suggesting that life is not a sudden leap in nature's evolution but a progressive step. Various combinations of elements create new characteristics which have 'feelings' with a varying degree of sensitivity. The CHON combination of carbon, hydrogen, oxygen, and nitrogen creates that characteristic having a degree of sensitivity we call life.

> In Chapter 3 of his book *On Purpose*, Charles Birch discusses what he calls "internal relations", and goes on to suggest that these internal relations are based on recent experience – implying memory. This is consistent with the concept of elemental characteristics. Since characteristic waveforms can be modified by association with other waveforms we have the creation of memory. Remember the homeopathic principle.

157

The homeopathic principle leads to a possible understanding of evolutionary change. If, as it appears to be the case, the introduction of a new aspect to a material characteristic becomes a permanent feature, this will apply to the material of living bodies. The blood stream is the conveyor of our fundamental characteristic which is carried in the genes; it circulates throughout the whole body – including the brain. Bearing in mind that all matter is fundamentally composed of electromagnetic waves which can, themselves, be influenced by the presence of electromagnetic waves, we can see that developed characteristics of the mind – those that have become our soul, will slowly and inexorably have an influence on the physical properties of the genes – which are also comprised of E.M.G. waves. In contrast to the views of Sigmund Freud, who believed that personality is fully formed in early childhood, evidence suggests the characteristic of a young adult will change as the mind further matures. With added experience of life, the strength of particular aspects will increase. This, in turn, could cause a thinking person to deviate, causing further change. As a result, the combination of parental genes which created the fundamental characteristic could now become modified. These thoughts lead to the possible support of Lamarckism.

> In 1809 Jean-Baptiste Lamarck proclaimed, in his Zoological Philosophy, that acquired traits could be inherited; however, I must add a significant caveat; we must take into account that the fully developed characteristic only matures late in life; whereas procreation – in humans usually takes place before middle age. Therefore the genes passed on will not be influenced by the fully formulated but by a developing mind.

Accepting *that* probability leads one to realise that non-thinking adults – those that simply follow, will be experiencing unfavourable development of their characteristic and consequently, their offspring will lack the benefits of evolution. This, in turn, leads to a sluggishness of certain sections of society. As is the case with regard to financial ability, the rich get richer while the poor get poorer, but in this case it is intellectual richness to which we are referring.

In his book "The Consolations of Philosophy", Alain de Botton draws attention to a pronouncement made by the renowned Roman philosopher Cicero during the century before the birth of Christ.

Due to its relevance, I repeat it here;

"There is no occupation so sweet as scholarship; scholarship is the means of making known to us, while still in this world, the infinity of matter, the immense grandeur of Nature, the heavens, the lands and the seas. Scholarship has taught us piety, moderation, greatness of heart; it snatches our souls from the darkness and shows them all things, the high and the low, the first and the last and everything in between; scholarship furnishes us with the means of living well and happily; it teaches us how to spend our lives without discontent and without vexation."

I think these words – and their full meaning, should be grasped by all students; it is the whole purpose of study and learning. Who might gain some benefit from these chapters, **certainly not those who do not read them?** One must think for oneself and re-educate oneself to make progress.

Chapter 16

Thinking of Science versus Religion
– in the light of the foregoing

Remembering that this whole hypothesis stemmed from my original thoughts regarding the nature of the electron, it isn't surprising that these thoughts led to other areas in the field of physics. However, it is quite surprising that it was impossible to avoid delving into spiritual matters. The links are there – regardless of one's religious views or objections to religious doctrines. If the pure energy manifest in electromagnetic-gravity waves *is* the basis of everything, then spiritual matters must be included.

At last those readers who gain moral support from prayer can see how their thoughts can link to the All-Enveloping-Influence and how focussing on that link will deflect mental interference by the many diverse, spurious signals. This is the essence of prayer; the focus of the mind to the unadulterated source, commonly referred to as God. However, this hypothesis also shows that one must learn to challenge the basis of myths and legends – no matter how this seems to conflict with traditional teachings. We must remember that when The Bible, The Koran, and other ancient scriptures were written, the majority of the world population was illiterate; the populous needed to absorb guidance through stories told as an analogy.

> Referring once again to Charles Birch's *On Purpose*, I am indebted to the author for quoting, as an introduction to his final chapter, the following words by the celebrated mathematician and philosopher A.N. Whitehead. "Progress in truth – truth of science and truth of religion – is mainly a progress in the framing of concepts, in discarding artificial abstractions or partial metaphors, and in evolving notions which strike more deeply into the root of reality."

> This is exactly what my hypothesis is all about, on the one hand discarding myth and legend while on the other hand, as with Asimov's *Tree of Science*, pruning off that which is proven to be false.
>
> The root of reality is that the EMG wave is the basis of everything.

The inclusion, in part one of this book, of a chapter entitled *Thinking of Science versus Religion – historically* was not intended as the theme of the whole hypothesis – merely as an exercise to help the reader view familiar matters from alternative viewpoints. It is, therefore interesting that, having started a mental investigation into sub-atomic 'particles' that we should find ourselves debating the association of science and religion. This, to me, indicates the fundamental nature of spiritually in all matters regardless of one's personal thoughts concerning the merits of any or all religions.

The universality of the EMG wave has also been shown to be fundamental; being the link between all physical matter – on an astronomical scale, or down to the minute, almost non-existent level of quantum physics. The evolvement of life and the function of highly developed brains depend upon these waves which also link our souls to our minds enabling us to live well and in harmony, whether or not we take advantage of that ability.

The hypothesis propounded in these pages *does* provide the concept within which *all* of the various aspects of universal existence can be brought together. It forms a basis for experts in each field to use the EMG wave as the link in their research to answer fundamental questions. If the General Unified Theory (GUT.) sought by physicists *is* found to be structured on the propagation and subsequent development of electromagnetic-gravity waves, and the neuroscientists can widen their holistic appraisal of the workings of the brain to accept this basis, we will truly have the focal point for all research. EMG waves being the ultimate reduction of *everything,* the complimentary, although contrary attributes of holism and reductionism have been brought together.

> As far back as 1907 Einstein declared that everything in the universe is a repository of enormous latent energy. He also contemplated that electromagnetism and gravity might have a common origin.
>
> [*Einstein* by Denis Brian]

The foregoing pages clearly show how this physical basis is also the root of spiritual understanding. Theologians can anchor their religious teachings on a fact; rather than having to rely on myth and legend. God becomes a tangible entity, albeit that of pure energy - truly the origin of everything. The strength of one's love of God is a measure of the compatibility of one's prime characteristic with the pure energy waves emanating from Creation or, better still, the re-creation. In other words, the less notice one takes of the spurious – and this includes the conflicting messages sent by the scriptures, the more focussed the mind will be towards the pure energy vibrations and the clearer will be one's thinking.

It is worth cogitating that, since the emanations from the creation of the universe are all enveloping and the origin of everything, any other form of life in the cosmos will evolve on the same basis. That's not to say that the physical form of any extraterrestrial being would be identical to the human but the fundamental wave-orientated structure would be similar. We must realise that Earth is just one planet of a star amongst millions in this galaxy which is amongst millions of galaxies in the universe; all of which are influenced by the same waves emanating from the same source – Creation, the Big-Bang. The sequence by which the materialistic elements developed do not **only** apply to us and our immediate world; the planets surrounding any star of any galaxy were created by the same physical process as was our own domain. All materials, whether dead or alive are subject to the same laws of nature; this is a fundamental truth. One might use the cliché that nature's laws are God's laws.

The concept that God has made nature's laws is often quoted as evidence that there is an overlord governing universal life. However, let us consider the fundamental sphere of mathematics. There are many rules put forward by ancient thinkers such as Pythagoras, with his right-angled triangle, Euclid with his geometric shapes, even Newton with his Calculus which might be construed to be God's word coming from human mouths. But, I suggest to you, that these laws, together with the much more sophisticated mathematical manipulative techniques employed to determine the answers to modern problems, are all developed from fundamentals which cannot be any other way.

For example; if I display my right index finger, unity is indicated. If I also display my left index finger, this also indicates unity – twice. Already we have made the discovery of addition and multiplication; unity plus unity can *only* equal the value of the numerical symbol we call 'two'. If we multiply unity in one position by two positions, we arrive at the same result. Addition and multiplication are shown to be closely related. Having mentioned Pythagoras let us look at his right-angled triangle. Anyone who has done any basic setting-out will be aware of the 3,4,5 triangle in which, to create a right-angled corner for a garden or work-place, sides of 3 units and 4 units together with a hypotenuse of 5 units, are measured - a right-angle will result. It also transpires that the area of that triangle is 6 units. Such mathematical laws are not the creation of man or God – they are fundamental. As already discussed, repeated applications of such simple steps allow the evolution of ultimately, very complex outcomes. Inversely, the structure of very complex bodies can be reduced to an ultimate simple base from which all else was built.

Thinking back; it does appear that *everything* can be reduced to electromagnetic-gravity waves - *but no further*. A chunk of rock has a molecular structure based on the atomic elements from which it is composed. These atomic elements have electron shells formed of EMG waves centred on nuclei of high energy EMG waves. Plants too have a similar structure but with a more sophisticated molecular composition capable of self development – life, due to the evolution of genes – which are also comprised of EMG waves. Animals are a more advanced form of life, having the addition of more DNA elements causing the development of a brain with a limited mind. Further additions to the DNA string has led to the development of the brain into a more sophisticated mind, capable of understanding language, triggering the evolution of mankind. This human species, the most highly developed on Earth is still evolving through the development of literature (communication), mathematics (logic), and music (harmony). Even our soul has been shown to be a combination of EMG waves, keying into those emanating from Creation.

> It is reported (by Richard Dawkins in *The God Delusion*), that Martin Luther was so worried about the intrusion of scientific logic into general knowledge that he declared "Reason is the greatest enemy that faith has ..."
>
> [This is surely because reason had not yet pursued the matter far enough. One *can* have faith within the bounds set out by reason – faith in the ultimate truth.]

It is not everyone that can analyse matters presented to them, but everyone should be prepared to *think*. As outlined at the very beginning of this book, thinking should become a habit; this is the only way to achieve understanding. Consider the analogy of the Sun versus the universe. The Sun radiates energy to the Earth. Nature utilises that energy in a multitude of ways – enhancing the development of mankind. The 'Sun-God's' energy pervades everything on Earth. The 'Creative God's' energy pervades even the Sun – and all the other bodies of the universe.

Even if the beliefs of religious adherents are based on myth and legend; lacking a coherent basis, one must take into account the placebo effect of appealing to the heavens for guidance. In the world of medicine, most sick people have little idea of actually how the prescribed medicine works but it still cures them – hopefully, whether or not the chemistry was responsible. This is a case of 'blind' belief; the patient believes in the expertise of the doctor. People do have spiritual beliefs even though they have not reasoned the logic.

Many non believers are also quite reluctant to reject God's possible existence altogether, they still secretly invoke spiritual support. This is quite reasonable in the context of this hypothesis; in fact they are fundamentally more correct in that they are, unknowingly, seeking the primary energy waves – the All-Enveloping-Influence, rather than some mythical god declaring an arbitrary set of rules for living.

> It is unfortunate that the great scientist, Richard Dawkins, has such an unyielding approach to the matter of science versus religion. This only serves to cause readers to dig their toes in. It appears that it

hasn't occurred to him that there is room for both. Believers must learn to accept natural selection while scientists must allow for personal beliefs - even if they are initially based on unsound teachings.

Richard Dawkins – as most other people, appear to make the mistake of thinking of God as a separate entity. God (universal energy – the All-Enveloping-Influence) is *literally* in everyone's mind - waiting to be discovered.

So it is that even the most sophisticated being, the human, is a direct evolvement from EMG waves. The difficulty in appreciating this as a fact is due to the great complexity - the remoteness of the end result from the irreducible origin.

Consider the analogy of a large centre of population. A great city has a character of its own based on the complex composition of buildings, infrastructure and, of course, its population. The way individuals go about their day-to-day lives adds colour to the character of the whole metropolis - even if each person makes such a small contribution. Animals, both large and small, also contribute to the overall character of the city domain; pets taking walks in the park, hungry dogs scavenging the rubbish bins of mean alleyways, pigeons pooping on the window-ledges of apartment blocks, perhaps mosquitoes breeding in open sewers before attacking the passer-by. Add to all this the kerb-side trees, flowers in suburban gardens and the beauty of well-laid-out parks and we see the complexity of detail which builds to create the character of the city. On top of all that potential variance we humans have language; language which itself continues to vary with different accents within different tongues causing the development of cultural enclaves. The human character evolves in much the same way: every incident whether large or small, good or bad, adjusts the accumulated wave-form of the individual mind. As this library of experience develops, the personality develops - for the better or for the worse; the whole city becomes a living cultural body.

It does seem rather trifling to reduce everything to a matter-of-fact fundamental base; surely life is more than just waves. You might ask; what about emotions, appreciation-of-art, love, and hate? As propounded earlier, all of these facets *do* have a wave foundation. For instance, what is the difference between a painting and a work of art? - Vibrations. An inspired artist transfers his emotions via his brush to the paint on the canvas. The work becomes imbued with his thoughts and passions; someone harmonising with the work will be similarly stimulated. Someone else, not in sympathy with the artist's thought waves, will merely see paint.

Let us just look once again at the analogy of the computer. We are increasingly aware of the wondrous tasks performed by this marvel of our age; look at the colours, the shapes, the sounds. Combined together, these attributes can, for example, be made to create fantastically realistic animations, difficult to differentiate from live filming. If we consider the basics of electronic computers, we find that they rely on nothing more than the movement of electrons. All that wonder is reduced to these tiny, sub-atomic 'particles' - but, even they are reducible to waves.

Each electron, each proton, each neutron, being formed of waves, can have character instilled within its vibrations.

> In an article entitled "In search of Origins", Gerald F. Colvin draws attention to Galileo's letter to the Grand Duchess Christina in which he stated "The Bible tells us how to go to heaven – not how the heavens go." He goes on to point out that the focus of the Book of Scripture is different from that of the book of nature, and should not be unduly pressed into the daily service of science. In contrast, Colvin observed that the great scientists Kepler and Newton regarded themselves as "priests of God in the temple of nature." More recently physicists Stephen Hawking and Paul Davies have spoken of uncovering "The Mind of God." It appears significant that, although science and religion to the casual observer seem to be poles apart, when one truly delves into fundamentals these views converge to focus on the same point of origin.

Thinking of Science Versus Religion – in the Light of the Foregoing

The considerations and thoughts referred to in the preceding pages, although basically dealing with physical fundamentals, have inevitably led us, also, to spiritual matters; this only serves to underline that the state of matter is linked, quite positively, with our state of mind, and our spiritual wellbeing; the link being the EMG wave.

To discover the magic of waves, with their unifying link, brings the possibility (albeit the unlikelihood) of universal harmony. We can see how, with training, consideration-for-others could become a well developed sector of our characteristics and possibly bring about a long-term solution to world conflict

Is this 'Pie in the Sky' or a realistic target? It is up to us all.

Part 4 – The Conclusion

Chapter 17

Thinking of the Explanation of Everything

It is very reasonable that we should all seek happiness, after all why be miserable if you **can** be happy. The question arises how **do** we achieve happiness? I believe **fulfilment** brings true happiness; success with ones endeavours, a harmonious family relationship, an existence free of 'conflictics', all lead to fulfilment. However, the pleasure of **understanding** surpasses all; it is far superior to financial wealth, it is much more long lasting than sex, or the adrenaline rush created by extreme sports. Just think of the potential for happiness, how fulfilled we would be, if we truly understood how everything keyed together.

> I am indebted to Marcelo Gleiser for drawing attention, in his book *The Dancing Universe*, to an ancient poem written in the 1st century B.C.E. by the Roman poet, **Lucretius**. Entitled *The Nature of the Universe* he hits the nail on the head with regard to God and Creation. The following is Gleiser's extract:
>
>> This dread and darkness of the mind cannot be dispelled by the sunbeams, the shining shafts of the day, but only by an understanding of the outward form and inner workings of nature. In tackling this theme, our starting point will be this principle: *Nothing can ever be created by divine power out of nothing.* The

reason why mortals are so gripped by fear is that they see all sorts of things happening on the Earth and in the sky with no discernible cause, and these they attribute to the will of God. Accordingly, when we have seen that nothing can be created out of nothing, we shall then have a clearer picture of the path ahead; the problem of how things are created and occasioned without the aid of gods.

Or, a little later:

For the mind wants to discover by reasoning what exists in the infinity of space that lies out there, beyond the ramparts of this world - that the region into which the intellect longs to peer and into which the free projection of the mind does actually extend its flight.

The thoughts set out in the pages of this book, *The Link, are* the potential portal through which the mind can project into the infinity of space or the improbability of quantum physics.

The logical approach to the creation of the universe and the origins of mankind is frequently criticised as being cold and matter-of-fact; not having the necessary spiritual core that inspires desirable moral codes and is devoid of emotions. The memory bank - the interaction of waves within the brain, enhanced by sights, sounds, smells, taste and touch, is what causes emotion. This may be physical fact but, as mentioned earlier, elements do have 'feelings' and these 'feelings' aggregate to create emotions. *An oxygen atom taken from water will vibrate slightly different from an oxygen atom taken from say sulphuric acid. It will be at the same frequency but the waveform will have a slightly different shape – it has remembered its previous experience.*

169

Theologians have been found guilty of building a regime based on myth and legend; many of the tenets having no factual basis; whereas the foregoing has clearly shown a strong spiritual link to the materialistic foundation laid down by the Creation.

All that has preceded the conclusion in this book is based on one physical fact – everything and everybody stems from a common source of pure energy. This surely is a comfort; it is a hook to hang everything on; an unadulterated source still accessible to everyone; a firm foundation. This energy has interacted and reacted to create all materials, some of which have further developed into living beings – culminating in modern mankind. Not only has this energy been the creator, or re-creator, of all things and all beings, it permeates the minds of us all and even creates the soul. Here is the spiritual truth.

At the risk of appearing too matter-of-fact, let us summarise the conclusions derived from this hypothesis.

The universe, being the complex development of electromagnetic-gravity waves in a multitude of forms throughout space, exists permanently but is ever changing. Since the change is cyclic, there is no beginning nor is there an end; we can only measure the current state by reference to a critical point in the cycle. The critical point nearest to our present state is the instant of the Big-Bang - when only pure energy existed; the point in the cycle when a negative universe had been compressed to zero, in a universal black hole, prior to the explosive rebirth into the universe we know today.

As the waves of energy expand from the re-creation a complex process of conversion to matter is set in motion by the evolvement of pressurised hot-spots. In the turmoil of electromagnetic-gravity (EMG) activity, protons, neutrons, and electrons are formed from transient sub atomic 'particles', progressively combining to create a variety of atoms. These atoms, initially hydrogen then helium, combine and reform into oxygen, carbon, nitrogen, iron, silver, gold, and all the other elements found in the cosmos. The combined gravitational forces draw these 'particles' together into great dust clouds from which stars are formed these, in turn, are gathered together into clusters of galaxies.

By this method, the EMG waves have been converted into what has been called 'particulate' matter (which I have renamed Deggs) at the expense of the pure energy. As this process develops the universe expands but, progressively the rate of expansion slows as the raw energy is consumed. Ultimately, the total gravitational forces of all the matter, and the yet-to-be-converted energy, will become greater than the inertia of the system, resulting in a reversal of the whole process. As the gravitational forces pull the whole universe together, the process will accelerate, causing an increase in temperature and pressure which will progressively convert matter back into pure energy – in another universal black hole (The Grand Wave, all as illustrated by Diagram 10 on page 111).

Ultimately, all the matter will be converted back into pure energy and the universe will, momentarily cease to exist. However, the act of compressing all the matter back into energy, at immeasurable temperatures, will cause another re-creation – another Big-Bang. The inertia of this mega-process will likely cause the new universe to be a negative version of that in which we live.

Now let us stretch our imagination and visualise this Grand Wave turned back on itself to form a wavy circle or, better still, a vibrating sphere; similar to a soap bubble. This takes us back to the beginning of this hypothesis where the electron is considered as a vibrating bubble, usually centred on the atomic nucleus. Now, with the grand wave in spherical form, all the positive energy of the current universe and the negative energy of the future (and previous) universes will be centred on the ultimate space – curved inwards.

From this we conclude that the universe is a continuous natural cycle not requiring any gods to set it on its way – it is the ultimate natural occurrence. *This, in essence, is what the adherents of the Indian religion of Jainism believe.* It is important to realise that this fundamental concept of a cyclic universe – not yet accepted by scientists, should act as a focus for men of science *and* men of the Church; it is just a matter of viewing a situation from different perspectives. Science and religion both revere nature.

We all should view the myths currently contained in religious scriptures as a purely temporary means of conveying the beliefs necessary to focus the human character. As science progressively establishes the truth of the natural environment, facts should be allowed to replace fiction.

In his book *The Dancing Universe,* Marcelo Gleiser makes reference to Thales of Miletus, the 6th century Greek philosopher who, by some roundabout means seems to instinctively have arrived at the core of this hypothesis. He searched for a unifying principle within nature itself – he even referred to magnetism in all things, going on to declare "all things have a soul", and "all things are full of gods". His foresight seems to have been dismissed because he chose water, or moisture, as the probable common factor; which now appears so unreasonable. Of course he did not have the benefit of modern understanding of atomic structure which was, I remind you, the starting point for your author's realisation that everything is ultimately reducible to electromagnetic-gravity waves. In turn, these have a common origin - the re-creative Big-Bang.

The pure energy, radiating from each re-creation is a common factor of everything and everybody. This then is what believers really mean when they invoke God – whether or not they understand the fundamentality of EMG waves, atomic theories, Big-Bangs, or re-creations. It is from these radiations of energy waves that all matter is made and ultimately from this all life is born. The creator of all then is the pure energy – *God; to those who wish to identify the supreme source.* We must also note that this pure energy – 'God', has made the whole universe – not just Earth.

The notion of the cyclic nature of pure energy expanding into matter, then contracting back to pure energy, answers the question as to where God came from – the pure energy which may be referred to as God, exists in perpetuity either as matter or as energy.

We also see from this concept that God does not direct one's life; each individual is free to choose his or her own direction. However, the all embracing emanations direct from the universal energy source provide a clear, harmonious characteristic as a guidance; a path to follow. Your personal tendency to accept, to ignore, or even reject that support depends upon your upbringing; how much of your characteristic is still based on the smooth, unadulterated wave-form from the recreation versus how much reliance is placed upon the spurious waves from the multitude of sources – close by or distant.

The realisation that the brain is internally an interactive organ, subject to inter-compartmental induction, leads to an understanding of personality development by the absorption of life's experience. The same is true of society as a whole; various factions throughout the world develop characteristics which come into contact with others. In some cases induction takes place causing the harmonious assimilation into a newly developed character. Alternatively, conflict occurs which leads to rejection. This rejection is probably simply due to the two characteristics being out-of-phase. If consideration would allow time to adjust the phase relationship, a harmonious merging could result. Of course the will to adjust must be there in the first place; this, in turn, depends upon the historic development of the existing characteristics.

If universal truth be allowed to override myth and legend, all factions would progressively focus on the same origin – we would, at last be in phase: have harmony. Alas, religious inertia is such that it took until 1992 for the Catholic Church to reverse the 1633 condemnation of Galileo – 359 years, just to accept one man's observations of true knowledge. However, maybe there is hope; this time-lag seems to be improving – it took something like 2,000 years to reject Aristotle's crystal spheres.

As your author read of the various attempts of the ancient Greek philosophers and mathematicians to unravel the mysteries of the cosmos, a common failing became apparent. Thales, followed by Pythagoras, Plato, Aristotle, Archimedes and Ptolemy, with a smattering of others in between, believed – instinctively, that there is some underlying force or energy creating the heavens and the Earth. In the more enlightened period of the renaissance, Copernicus, Brahe, Kepler, and, subsequently, even Newton continued the development of cosmological theories – they too seemed aware of an underlying commonality but, apart from discounting Divine influence for a number of reasons, none were able to find the key.

> Would it be too presumptuous of me, your author, to declare that I **have** found the key, or rather the link – with the aid of modern knowledge and an open mind – uncluttered by scientific purpose?

End Note

The hypothesis presented here has not been prepared by a highly qualified scientist and is likely to be treated by them with much scepticism. Theologians will probably tend to reject the matter-of-fact treatment of spiritual matters which deflects the mystery of simple belief, but if truth is found in these pages one should not dismiss the relevance. However, readers, whether qualified or not, should consider the content with an open mind - this is, in itself, the first principle of good science. Your author hopes that the common link portrayed above **will** be seriously considered by fundamental scientists, mathematicians and theologians **- as a parallel structure,** in the belief that EMG waves will be truly found to be the basis of everything and the ultimate focus of all research.

When one considers the tremendous progress that has been made in recent decades by scientists to discover fundamental truths, and by engineers and technicians of all disciplines to make use of that knowledge, it is not possible to envisage to where that will lead us in the next hundred or so generations. Knowledge is growing exponentially.

Is it conceivable that mankind's ultimate purpose is to modify the development of the universe so that future generations may reap the fruits of our endeavours in an everlasting environment rather than being swallowed by an exploding star or disappearing in a great black hole?

> In that connection, I must repeat a quotation found in Charles Birch's *On Purpose.*
>
>> "What happens really matters if it matters ultimately, and it matters ultimately only if it matters everlastingly.

> What happens can matter everlastingly if it matters to him who is everlasting.
> Hence, seriousness about life implicitly involves faith in God."
> John B. Cobb, *God and the World.* (1959)

[If mankind is to be everlasting, then we do have a purpose – to make it so.]

It is worth pointing out that this hypothesis was initially prepared without direct reference to any other work; the thought train being based solely on the author's general knowledge and varied personal experiences over seven decades. The whole matter was then researched; discovering much supporting material but also some conflicting work, to both of which references have been made. These references merely serve as illustrations to show that although this original concept may be considered radical, in the past others have touched on the theme even though the fundamental base was not considered.

The object here, in these pages, is to draw attention to the commonality of the ultimate reduction of all subject matter. Although no attempt has been made to **prove** any of the above, the circumstantial evidence presented is overwhelming. It is up to the reader to apply his/her own thinking ability to determine its worth.

> Sigmund Freud has been quoted as saying "It is a mistake to believe that a science consists in nothing but conclusively proved propositions, and it is unjust to demand that it should."

Let us assume the truth of the foregoing thesis until conclusively proven to be otherwise.

Appendices

A LOGIC PUZZLE

The ages of Anne and Bill summate to 91 years.
Anne is three times the age Bill was when Anne was
twice as old as Bill is now.
What are the ages of Anne and Bill now?

SOLUTION

If you find yourself going around in circles, think pictorially.

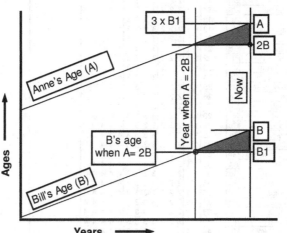

We are given that (1) A + B = 91 and that (2) A = 3 x B1
From the diagram we see that (3) A - 2B = B - B1
Subs. (2) in (3) 3B1 - 2B = B - B1 & 4B1 = 3B & B1 = 3B/4
From the diagram B - B1 = B/4 & (4) A = 9B/4
Subs.(4) in (1) 9B/4 + B = 91 & 13B/4 = 91 & B = 364/13 = 28
91 - 28 = A = 63

ANNE IS 63 AND BILL IS 28

177

Electron Shell Structure per M.G.Duffill (p1)

Element		Z	1sx2	2sx3	3sx4	4sx5	5sx6	6sx7
Shells Full			2	8	20	40	70	112
Max/Shell			2	6	12	20	30	42
Hydrogen	H	1	1					
Helium	He	2	1 1					
Lithium	Li	3	1	2				
Beryllium	Be	4	1 1	2				
Boron	B	5	1	2 2				
Carbon	C	6	1 1	2 2				
Nitrogen	N	7	1	2 2 2				
Oxygen	O	8	1 1	2 2 2				
Flourine	F	9	1 1	2 2	3			
Neon	Ne	10	1	2 2 2	3			
Sodium	Na	11	1 1	2 2 2	3			
Magnesium	Mg	12	1 1	2 2	3 3			
Aluminium	Al	13	1	2 2 2	3 3			
Silicon	Si	14	1 1	2 2 2	3 3			
Phosphorus	P	15	1 1	2 2	3 3 3			
Sulphur	S	16	1	2 2 2	3 3 3			
Chlorine	Cl	17	1 1	2 2 2	3 3 3			
Argon	Ar	18	1 1	2 2	3 3 3 3			
Potassium	K	19	1	2 2 2	3 3 3 3			
Calcium	Ca	20	1 1	2 2 2	3 3 3 3			
Scandium	Sc	21	1 1	2 2 2	3 3 3	4		
Titanium	Ti	22	1 1	2 2	3 3 3 3	4		
Vanadium	V	23	1	2 2 2	3 3 3 3	4		
Chromium	Cr	24	1 1	2 2 2	3 3 3 3	4		
Manganese	Mn	25	1 1	2 2 2	3 3 3	4 4		
Iron	Fe	26	1 1	2 2	3 3 3 3	4 4		
Cobalt	Co	27	1	2 2 2	3 3 3 3	4 4		
Nickel	Ni	28	1 1	2 2 2	3 3 3 3	4 4		
Copper	Cu	29	1 1	2 2 2	3 3 3	4 4 4		
Zinc	Zn	30	1 1	2 2	3 3 3 3	4 4 4		
Gallium	Ga	31	1	2 2 2	3 3 3 3	4 4 4		
Germanium	Ge	32	1 1	2 2 2	3 3 3 3	4 4 4		
Arsenic	As	33	1 1	2 2 2	3 3 3	4 4 4 4		
Selenium	Se	34	1 1	2 2	3 3 3 3	4 4 4 4		
Bromine	Br	35	1	2 2 2	3 3 3 3	4 4 4 4		
Krypton	Kr	36	1 1	2 2 2	3 3 3 3	4 4 4 4		
Rubidium	Rb	37	1 1	2 2 2	3 3 3	4 4 4 4 4		
Strontium	Sr	38	1 1	2 2	3 3 3 3	4 4 4 4 4		
Yttrium	Y	39	1	2 2 2	3 3 3 3	4 4 4 4 4		
Zirconium	Zr	40	1 1	2 2 2	3 3 3 3	4 4 4 4 4		
Niobium	Nb	41	1 1	2 2 2	3 3 3 3	4 4 4 4	5	
Molybdenum	Mo	42	1 1	2 2 2	3 3 3	4 4 4 4 4	5	
Technetium	Tc	43	1 1	2 2	3 3 3 3	4 4 4 4 4	5	
Ruthenium	Ru	44	1	2 2 2	3 3 3 3	4 4 4 4 4	5	
Rhodium	Rh	45	1 1	2 2 2	3 3 3 3	4 4 4 4 4	5	
Palladium	Pd	46	1 1	2 2 2	3 3 3 3	4 4 4 4	5 5	
Silver	Ag	47	1 1	2 2 2	3 3 3	4 4 4 4 4	5 5	
Cadmium	Cd	48	1 1	2 2	3 3 3 3	4 4 4 4 4	5 5	
Indium	In	49	1	2 2 2	3 3 3 3	4 4 4 4 4	5 5	
Tin	Sn	50	1 1	2 2 2	3 3 3 3	4 4 4 4 4	5 5	
Antimony	Sb	51	1 1	2 2 2	3 3 3 3	4 4 4 4	5 5 5	
Tellurium	Te	52	1 1	2 2 2	3 3 3	4 4 4 4 4	5 5 5	
Iodine	I	53	1 1	2 2	3 3 3 3	4 4 4 4 4	5 5 5	
Xenon	Xe	54	1	2 2 2	3 3 3 3	4 4 4 4 4	5 5 5	
Caesium	Cs	55	1 1	2 2 2	3 3 3 3	4 4 4 4 4	5 5 5	
Barium	Ba	56	1 1	2 2 2	3 3 3 3	4 4 4	5 5 5 5	

Electron Shell Structure per M.G.Duffill (p2)

Shells Full			2	8	20	40	70	112
Max/Shell			2	6	12	20	30	42
Element		Z	1sx2	2sx3	3sx4	4sx5	5sx6	6sx7
Lanthanium	La	57	1 1	2 2 2	3 3 3	4 4 4 4	5 5 5	
Cerium	Ce	58	1 1	2 2	3 3 3 3	4 4 4 4	5 5 5	
Praseodymium	Pr	59	1	2 2 2	3 3 3 3	4 4 4 4	5 5 5	
Neodymium	Nd	60	1 1	2 2 2	3 3 3 3	4 4 4 4	5 5 5	
Promethium	Pm	61	1 1	2 2 2	3 3 3 3	4 4 4	5 5 5 5	
Samarium	Sm	62	1 1	2 2 2	3 3 3	4 4 4 4	5 5 5 5	
Europium	Eu	63	1 1	2 2	3 3 3 3	4 4 4 4	5 5 5 5	
Gadolinium	Gd	64	1	2 2 2	3 3 3 3	4 4 4 4	5 5 5 5	
Terbium	Tb	65	1 1	2 2 2	3 3 3 3	4 4 4 4	5 5 5 5	
Dysprosium	Dy	66	1 1	2 2 2	3 3 3 3	4 4 4	5 5 5 5 5	
Holmium	Ho	67	1 1	2 2 2	3 3 3	4 4 4 4	5 5 5 5 5	
Erbium	Er	68	1 1	2 2	3 3 3 3	4 4 4 4	5 5 5 5 5	
Thulium	Tm	69	1	2 2 2	3 3 3 3	4 4 4 4	5 5 5 5 5	
Ytterbium	Yb	70	1 1	2 2 2	3 3 3 3	4 4 4 4	5 5 5 5 5	
Lutetium	Lu	71	1 1	2 2 2	3 3 3 3	4 4 4 4	5 5 5 5	6
Hafnium	Hf	72	1 1	2 2 2	3 3 3 3	4 4 4	5 5 5 5 5	6
Tantalum	Ta	73	1 1	2 2 2	3 3 3	4 4 4 4	5 5 5 5 5	6
Tungsten	W	74	1 1	2 2	3 3 3 3	4 4 4 4	5 5 5 5 5	6
Rhenium	Re	75	1	2 2 2	3 3 3 3	4 4 4 4	5 5 5 5 5	6
Osmium	Os	76	1 1	2 2 2	3 3 3 3	4 4 4 4	5 5 5 5 5	6
Iridium	Ir	77	1 1	2 2 2	3 3 3 3	4 4 4 4	5 5 5 5	6 6
Platinum	Pt	78	1 1	2 2 2	3 3 3 3	4 4 4	5 5 5 5 5	6 6
Gold	Au	79	1 1	2 2 2	3 3 3	4 4 4 4	5 5 5 5 5	6 6
Mercury	Hg	80	1 1	2 2	3 3 3 3	4 4 4 4	5 5 5 5 5	6 6
Thallium	Tl	81	1	2 2 2	3 3 3 3	4 4 4 4	5 5 5 5 5	6 6
Lead	Pb	82	1 1	2 2 2	3 3 3 3	4 4 4 4	5 5 5 5 5	6 6
Bismuth	Bi	83	1 1	2 2 2	3 3 3 3	4 4 4 4	5 5 5 5	6 6 6
Polonium	Po	84	1 1	2 2 2	3 3 3 3	4 4 4	5 5 5 5 5	6 6 6
Astatine	At	85	1 1	2 2 2	3 3 3	4 4 4 4	5 5 5 5 5	6 6 6
Radon	Rn	86	1 1	2 2	3 3 3 3	4 4 4 4	5 5 5 5 5	6 6 6
Francium	Fr	87	1	2 2 2	3 3 3 3	4 4 4 4	5 5 5 5 5	6 6 6
Radium	Ra	88	1 1	2 2 2	3 3 3 3	4 4 4 4	5 5 5 5 5	6 6 6
Actinium	Ac	89	1 1	2 2 2	3 3 3 3	4 4 4 4	5 5 5 5	6 6 6 6
Thorium	Th	90	1 1	2 2 2	3 3 3 3	4 4 4	5 5 5 5 5	6 6 6 6
Protoactinium	Pa	91	1 1	2 2 2	3 3 3	4 4 4 4	5 5 5 5 5	6 6 6 6
Uranium	U	92	1 1	2 2	3 3 3 3	4 4 4 4	5 5 5 5 5	6 6 6 6
Neptunium	Np	93	1	2 2 2	3 3 3 3	4 4 4 4	5 5 5 5 5	6 6 6 6
Plutonium	Pu	94	1 1	2 2 2	3 3 3 3	4 4 4 4	5 5 5 5 5	6 6 6 6
Americium	Am	95	1 1	2 2 2	3 3 3 3	4 4 4 4	5 5 5 5	6 6 6 6 6
Curium	Cm	96	1 1	2 2 2	3 3 3 3	4 4 4	5 5 5 5 5	6 6 6 6 6
Berkelium	Bk	97	1 1	2 2 2	3 3 3	4 4 4 4	5 5 5 5 5	6 6 6 6 6
Californium	Cf	98	1 1	2 2	3 3 3 3	4 4 4 4	5 5 5 5 5	6 6 6 6 6
Einsteinium	Es	99	1	2 2 2	3 3 3 3	4 4 4 4	5 5 5 5 5	6 6 6 6 6
Fermium	Fm	100	1 1	2 2 2	3 3 3 3	4 4 4 4	5 5 5 5 5	6 6 6 6 6
Menelevium	Md	101	1 1	2 2 2	3 3 3 3	4 4 4 4	5 5 5 5	6 6 6 6 6 6
Nobelium	No	102	1 1	2 2 2	3 3 3 3	4 4 4	5 5 5 5 5	6 6 6 6 6 6
Lawrencium	Lr	103	1 1	2 2 2	3 3 3	4 4 4 4	5 5 5 5 5	6 6 6 6 6 6
		104	1 1	2 2	3 3 3 3	4 4 4 4	5 5 5 5 5	6 6 6 6 6 6
		105	1	2 2 2	3 3 3 3	4 4 4 4	5 5 5 5 5	6 6 6 6 6 6
		106	1 1	2 2 2	3 3 3 3	4 4 4 4	5 5 5 5	6 6 6 6 6 6 6
		107	1 1	2 2 2	3 3 3 3	4 4 4 4	5 5 5 5 5	6 6 6 6 6 6 6
		108	1 1	2 2 2	3 3 3 3	4 4 4	5 5 5 5 5	6 6 6 6 6 6 6
		109	1 1	2 2 2	3 3 3	4 4 4 4	5 5 5 5 5	6 6 6 6 6 6 6
		110	1 1	2 2	3 3 3 3	4 4 4 4	5 5 5 5 5	6 6 6 6 6 6 6

Select Bibliography

Barrow, John D. – *Theories of Everything* - B.C.A. 1991
The quest for ultimate explanation

Barrow, John D. – *Pi in the Sky* – Clarendon Press, Oxford 1992
Counting, thinking and being

Beaumont, J. Graham – *Brain Power* – Andromeda Oxford Ltd 1989

Birch, Charles –*On Purpose* – New South Wales University Press 1990
A consideration of meaning and purpose.

Calder, N. – *Einstein's Universe* – Penguin Books, N.Y. 1981
Based on TV series

Davies, Paul – *Are we alone?* – Penguin 1995
Implications of discovering extraterrestrial beings

Davies, Paul – *God and the New Physics* – Penguin 1990
God through science

Davis, P.C.W. – *The Search for Gravity Waves* – Cambridge University
Press 1980
A discussion of the nature of gravity waves

Dawkins, Richard – *The God Delusion* – Bantam Press 2006
An assertion of the irrationality of belief in God and   resultant
grievous harm.

Devlin, Steven – *The Maths Gene* – Weidenfeld & Nicolson 2000
Explanation of why everyone has the ability to do
mathematics

Dyson, Freeman J. – *The Sun, The Genome, and The Internet* – Oxford
Univ. Press 1999
Thinking of the future

Ferris, Timothy – *World Treasury of Physics, Astron. and Maths* – Little
Brown and Co 1991
A collection of articles related to science physics and
mathematics.

Feynman, Richard P. – *The Pleasure of Finding Things Out*
– Penguin Books 2000
A peep into the mind of one of the world's great scientists –
one who thinks

Garner, M. – *The Relativity Explosion* – Vintage, N.Y. 1976
A revision of the 1962 classic *Relativity for the Millions*

Gleick, James – *Genius, Richard Feynman and modern physics* – Little,
Brown & Co 1992
An excellent review of complex subject matter.

Gleick, James – *Chaos* – Vintage 1998
A very readable account of the mathematical sphere
known as Chaos

Greenfield, Susan – *The Private Life of the Brain* – Penguin Books 2000
A detailed look at the workings of the brain

Gribbin, John – *In Search of the Edge of Time* – B.C.A. 1992
Black holes and hyperspace connections

Gribbin, John – *In Search of Schrodinger's Cat* – Black Swan 1992
Quantum physics and reality

Hawking, Stephen – *A Brief History of Time* – Bantam Press 1991
A review for the layman of cosmological theories

Hawking, Stephen – *The Universe in a Nutshell* – Bantam Press 2001
An update covering the breakthroughs since 1991

Hellemans & Brunch – *The Timetables of Science* – Simon & Schuster 1988
A chronology of scientific history

The Link – from before big bang to brain and beyond

Krauss, Lawrence, M. – *Atom* – Little, Brown & Co 2001
    An odyssey from the Big-Bang to life on Earth

Magee, Bryan – *The story of Philosophy* – Dorling Kindersley Limited 1998
    An excellent review of the work of the famous philosophers.

Mermin, N.D. – *Space and Time in Special Relativity* – McGraw-Hill, N.Y. 1968
    Elementary exposition (some algebra)

Pais, A – *Subtle is the Lord* – Oxford University Press, N.Y. 1980
    The science and life of Albert Einstein

Pinker, Steven – *The Language Instinct* – Penguin Books 1995
    The origins and nature of language

Taylor, John – *When the Clock Struck Zero* – Picador 1994
    Big-Bang to brain, mind and consciousness

Van Over, Raymond – *Unfinished Man* – World Publishing 1972
    The human mind from the view of a parapsychologist

Watch Tower – *Life-How did it get here?*
    – Watch Tower Bible & Tract Soc. of Penns. 1985
    Discussion of Creation versus Evolution

Will, M. Clifford – *Was Einstein Right?* – Oxford University Press 1988
    Putting general relativity to the test

Zimmer, Carl – *Evolution, The Triumph of an Idea* – Harper Collins 2001
    Darwin's theory reviewed and explained.

Index

1.5 to 2 billion years ago., 45
12 billion years after creation, 73
12 billion years ago, 35
15th century, 13
16th century, 14, 38, 145
17th century, 39
18th century, 13, 57
1920s, 39
1940s, 39
20th century, 39, 48, 102
2nd century, 38
4.5 billion years ago, 45
4th century BCE, 12
5,000 million years ago, 44
500 million years ago, 45
6th century, 171

A

abstract ideas, 23
acceleration, 34
acceptance or rejection, 66, 131, 132
acquired knowledge, 9
adulthood, 68
algae type life, 45
Allah, 87
All-Enveloping-Influence, 18, 37,
    64, 65, 67, 69, 72, 111, 115, 116,
    124, 135, 136, 160, 164, 165
alphabet of life, 44
Americanism, 148
amino acids, 43, 44, 45
analogue to digital converter, 63
animal, 45, 60, 64, 92, 94, 114, 121,
    138
animal instinct', 64

animals, 60, 64, 121, 163, 165
antiparticles, 102
archaeologists, 47
Archimedes, 173
Aristotle, 12, 13, 38, 173
art, 10, 24, 25, 57, 63, 77, 83, 95,
    121, 165
Asimov, Isaac, 17
Asimov's Tree of Science, 161
asteroids, 67
astrology, 91
at one, 90
atom, 10, 41, 43, 51, 97, 98, 102,
    103, 127, 157, 170
atom bombs, 29
aura, 63, 65, 67, 68, 69, 78, 79, 80,
    81, 83, 84, 88, 89, 117, 123, 124,
    126, 130, 135, 136, 137, 140,
    143, 144, 148, 152
awareness of self, 7, 66
axons and synapses, 10

B

Bach, 57
Bacon, Francis, 15
base for *everything*, 10, 116
Beethoven, 57
Benveniste, Dr Jacques, 123
Beryllium, 100
Bible, 13, 16, 50, 86, 160, 166
Big-Bang, 35, 37, 38, 39, 40, 43, 64,
    97, 104, 109, 114, 128, 155, 162
Birch, Charles, 157, 160, 174
birth, 18, 40, 62, 65, 67, 73, 87, 88,
    92, 104, 120, 128, 137, 141, 159
birth of Christ, 19

birth of science, 14
black hole, 15, 34, 102, 104, 170, 171, 174
Blackmore, Susan, 71
Botton, Alain de, 69, 159
Brahe, Tycho, 12, 173
brain, 32, 46, 53, 58, 59, 60, 61, 62, 63, 64, 65, 67, 68, 70, 74, 75, 78, 108, 109, 113, 115, 119, 129, 131, 132, 133, 137, 138, 141, 148, 150, 151, 152, 153, 156, 157, 158, 161, 163, 169, 173
brain cells, 62, 66
brain waves, 88
brainwaves, 58
Brian, Denis, 162

C

Caesium, *29, 33*
Calcium, 100
calendars and clocks, *31*
cancer virus, 49
Carbon, 42, 44, 93, 100, 157, 170
Catholic, 135, 139, 143
chaos, *23*
characteristic, 53, 63, 64, 65, 66, 67, 68, 69, 70, 72, 73, 74, 77, 78, 79, 80, 81, 82, 84, 86, 87, 88, 89, 90, 91, 108, 109, 110, 111, 112, 118, 119, 120, 121, 122, 123, 124, 125, 126, 129, 130, 131, 132, 133, 134, 135, 137, 138, 139, 140, 141, 142, 143, 145, 146, 147, 148, 149, 150, 151, 152, 153, 154, 158, 162
characteristic waveform, 66, 73
Cicero, 159

child, 8, 64, 65, 67, 69, 70, 74, 109, 111, 121, 122, 132, 138, 140, 141, 142, 144, 148
child's character, 65
CHON, 44, 157
Christ, 14, 27, 159
Christian, 134, 143
Christian era, 14
Christian school, 86
Cicero, 159
clairvoyants, 112
clocks, 31, 33
Cobb, John B, 174
collective aura, 82
collective unconscious, 78
collective waves, 136
Colvin, Gerald F, 166.
comets, 44, 67
communion, 25, 90, 111, 117
compatibility, 70, 162
complex waveforms, 67
compound characteristic, 108, 125
conception, 64, 65, 73, 115, 149
conflict, 13, 16, 25, 32, 65, 66, 68, 69, 72, 74, 79, 82, 88, 118, 125, 127, 131, 132, 134, 138, 139, 142, 160, 167, 173
conflictic, 65, 72, 87, 89, 126, 138, 139, 141, 142, 147, 151, 152
conscience, 88, 89, 125
consciousness, 7, 45
consideration, 20, 25, 35, 39, 72, 97, 118, 135, 136, 138, 149
consideration for others, 71, 74, 117, 139, 167
considering extremes, *25*
conversion of mass to energy, 39
Copernicus, 14, 39, 94, 173
cosmic egg, 102
cosmic events, 67

cosmic radiations, 73
cosmologists, 13, 15, 18
cosmology, 6
cosmos, 10, 15, 25, 39, 40, 41, 46,
   63, 64, 73, 75, 91, 92, 109, 114,
   117, 119, 162, 170, 173
creation, 14, 15, 35, 38, 40, 43, 45,
   64, 86, 92, 98, 104, 116, 121,
   127, 128, 135, 144, 157, 162,
   163, 169, 170, 171, 172
Creation, 51, 64, 90, 104, 109, 111,
   115, 116, 121, 124, 125, 135,
   162, 163, 168
creationism, 86
Creationist, 13, 86
crime waves, 82
criminal, 81, 87, 88, 118, 140
Critical Path Analysis, 22
cultural, 117, 134, 140, 143, 144,
   145, 146, 147, 149, 150, 165
cultures, 9, 134, 139, 143, 144, 146,
   147
curved space, 34
dance, *25*

D

Darwin, 13, 14, 45, 60, 86, 134
Darwin's "Origin of the Species", 13
Davies, Paul, 166
**Dawkins, Richard, 71, 87, 116,
   164- 165**
deaths, 44, 73
DEGGS, 4, 97, 102
Deuterium, 40
developed sub-characteristics', 71
developing child, 74
Devine Design, 44
difference between 'speed' and
   'velocity', 34

digital computer, *23*, *49*
distorted depth of focus, 35
Divine Design, 46, 48
DNA, 60, 75, 82, 108, 137, 163
dogma, 13, 25, 135
doppler effect, 33
Doppler, Christian, 33
dream, 7, 110
dreams, 110
drugs, 110, 149, 150, 151
Druids, 14
Duffill's eggs, 102
Duffron, 28, 29, 31
Dyson, Freeman, 133, 188

E

E=MC², 39, 127
Earth, 8, 12, 14, 27, 28, 29, 31, 38,
   43, 44, 46, 47, 49, 53, 114, 116,
   162, 164, 169, 172, 173
Earthlings, 29, 31, 46
EGGs, 101
Einstein, 14, 16, 31, 32, 33, 34, 39,
   94, **95-96**, 101, 102, 104, **106**,
   127, **162**
electrical pulses, 65
electromagnetic, 29, 30, 34, 35, 40,
   43, 51, 57, 58, 61, 62, 64, 65, 66,
   67, 73, 74, 75, 78, 89, 92, 97, 98,
   102, 103, 104, 108, 110, 111,
   113, 114, 121, 122, 123, 126,
   127, 128, 131, 137, 153, 155,
   156, 158, 160, 161, 163, 170, 172
electromagnetic waves, 29, 35, 51,
   58, 62, 75, 92, 102, 123, 155,
   156, 158
electromagnetic-gravity waves, *35*,
   98, 114, *115*, *137*

Electromagnetic-Gravity-Globule -
an 'EGG', 101
electron, 41, 45, 51, 98, 99, 100,
101, 102, 103, 104, 157, 160,
163, 166
electrons, 41, 42, 97, 98, 99, 100,
101, 103, 104, 114, 123, 127,
166, 170
embryonic, 41, 61, 62, 72, 73
EMG waves, 35, 63, 108, 137, 161,
163
emotions, 78, 83, 85, 118, 126, 144,
165, 169
energy into mass, 39
enquiring mind, 7
environment, 9, 10, 26, 28, 36, 37,
41, 59, 65, 72, 81, 82, 86, 87, 88,
89, 90, 106, 119, 122, 130, 133,
134, 136, 138, 140, 141, 142,
145, 147, 152, 171, 174
environment of our upbringing, 88
escape velocity, 34
ethereal, 67, 68, 73, 74, 80, 89, 90,
109, 110, 113, 115, 116, 119,
120, 124, 135
ethereal waveform, 73
ethereal waves, 67, 75, 109, 113,
115, 120, 125
Euclid, 27, 16 2
evolution, 37, 44, 45, 48, 49, 86, 87,
114, 134, 157, 158, 163
evolutionism, 86
evolutionists, 13
exercise in thinking, 36
expanding universe, 10, 128
expansion of the universe, 38, 104
experience, 8, 22, 25, 32, 59, 62, 66,
67, 68, 69, 70, 72, 74, 79, 80, 86,
89, 90, 91, 92, 95, 110, 119, 124,
125, 126, 130, 131, 132, 133,
134, 135, 136, 138, 139, 140,
141, 143, 144, 145, 146, 148,
149, 150, 151, 152, 154, 155,
157, 158, 173
extend one's knowledge, 22
Extra Sensory Perception, 113

F

faith, 87, 88, 94, 117, 143, 164, 175
father, 65, 67, 68, 133, 141
feeling, 83, 96, 136, 144
flexibility of mind, 9
foetus, 62, 64, 88
followers of fashion, 83
fossil finds, 45, 47
foster parents, 67
foundation, 10, 16, 17, 39, 57, 74,
75, 90, 97, 124, 140, 147, 165,
170
fractals, 23
frequencies, 53, 55, 56, 58, 67, 68,
92, 103, 120
frequency, 25, 29, 33, 53, 54, 55, 56,
58, 63, 64, 68, 69, 80, 88, 91,
101, 106, 115, 117, 118, 123, 155
Freud, Sigmund, 158, **175**
Friedman, Alexander, *39*
fundamental, 10, 18, 27, 33, 35, 40,
50, 55, 56, 57, 68, 69, 70, 72, 74,
87, 89, 90, 91, 95, 97, 103, 110,
117, 118, 119, 121, 124, 125,
126, 134, 140, 141, 143, 146,
158, 161, 162, 163, 165, 171,
174, 175
fundamental characteristic, 125, 158
fundamental scientists, 39
fundamental wave, 74

G

galaxies, 10, 15, 35, 73, 75, 92, 104, 162, 170
galaxy, 15, 75, 104, 162
Galileo, 14, **166,** 173
Gamma Rays, 58
gamma waves, 155
Gamov, George, 39
gene, 60, 61, 65, 70
gene theory, 70
general awareness, 10, 97
General Unified Theory, 161
genes, 13, 60, 61, 62, 64, 65, 66, 70, 158, 163
Genesis, 38, 86
Gleiser, Marcelo, 95-96, 168-169, 171-172
globules, 10
God, 7, 15, 18, 19, 37, 38, 46, 50, 75, 79, 80, 84, 86, 90, 111, 114, **115,** 116, 117, 125, 135, 136, 162, 164, 165, 166
going-with-the-flow', 83
Grand Unified Theory, 97
graphical pictures, 24
gravitational energy, 35
gravitational force, 102
gravity, 8, 30, 34, 35, 42, 44, 102, 103, 104, 106, 114, 121, 123, 126, 127, 128, 137, 160, 161, 162, 163, 170, 172
Gravity, 34, 104, 127, 128
gravity curves space, 34
Greeks, 12, 146
Greenfield, Susan, 151

H

habit of thinking, 9

Haeckel's "History of Creation", 13
harmonies, *25, 53, 56*
harmonious environment, 68, 142
harmonious vibrations, 70
harmony, 53, 56, 65, 69, 70, 71, 72, 77, 79, 85, 87, 90, 96, 108, 109, 121, 130, 131, 135, 139, 141, 142, 146, 147, 150, 152, 153, 155, 161, 163, 167, 173
Harmony, 53, 108
Hawking, Stephen, 166
heaven, 12, 13, 119, 125, 166
heavens, 46, 75, 159, 164, 166, 173
Heisenberg, 106
helium, 41, 42, 170
Helium, 44, 99
hell, 125
high speed flight, 8
Himmler, 81
history, 12
History, 18, 160
Hitler, 81, 82, 134
holiness, 136
holistic view, 10
Holocaust, 81
homeliness, 136
homeopathic principle, 157, 158
homeopathy, 109, 122, 123
homo-sapiens, 45
horologists, *28*
hot vents, 49
How and when will it come to an end?, 15
How can God be *in* all of us?, 125
How can He be 'in' us all?, 15
How can we access this source of spiritual strength?, 75
How did it all begin?, 15, 128
How did mankind develop, 15
how *does* the brain *know*?, 86

how God can be 'in' us all, 7
How has life evolved?, 40
how has the universe developed?, 40
how much control He exerts upon
    us, 7
Hubble, Edwin, 39
human, 6, 10, 37, 45, 57, 58, 59, 61,
    64, 66, 70, 73, 74, 80, 92, 108,
    114, 121, 129, 133, 135, 137,
    138, 149, 150, 151, 156, 162,
    163, 165, 171
human brain, 7, 58, 61, 64, 66, 74,
    80
human existence, 6
human experience, 6, 18
human relationships, 10
Hume, 48, 66
Humour, 132
Huygens, Christian, 98
hydrogen, 40, 41, 42, 99, 157, 170
Hydrogen, 44
hysteresis, 132

I

idea of thinking, 6
imagination, 7, 9, 11, 110, 156
inanimate materials, 10, 135, 140
induction, 62, 66, 132, 173
information from pictures, *23*
infrared waves, 58
in-phase, 91
inspiration, 63, 83, 90, 96
Is life predetermined?, 7, 126
Is there a God?, 7, 15, 124

J

Jainism, 171
Jewish race, 81

Jews, 81, 134, 139
Judaism, 14
Jung, Carl, 78, 110-111

K

Kamikaze pilots, 119
Kasparov, 133
keeping an open mind, 10
Kepler, 166, 173
Koran, 160
Krauss, Lawrence M, 37, 40/43, 45

L

**Lamarck, Jean-Baptiste de, 14,
    158**
Lamarckism, 158
language of the brain, 61
laser light, *31*
latent radiation., 39
laws of physics, 7, 62
learning skills, 60
Lemaitre, Abbe Georges, 39, 102
Leonard, Richard Anthony, 57
Levy, David H., 44
library of experience, 9, 165
life, 11, 19, 29, 30, 32, 34, 40, 43,
    45, 46, 47, 48, 49, 58, 65, 66, 68,
    70, 71, 72, 88, 92, 97, 106, 107,
    109, 117, 118, 119, 120, 122,
    124, 125, 126, 130, 133, 134,
    138, 140, 142, 143, 145, 147,
    149, 150, 151, 153, 154, 157,
    158, 162, 163, 165, 172, 173, 175
life after death, 6
life threatening situations, 151
life' after death, 10
life's puzzles, 7

light, 28, 29, 30, 33, 34, 35, 58, 62,
    84, 98, 102
light-speed, 62
literature, 76, 83, 122, 163
Lithium, 100
living bodies, 10, 60, 93, 120, 158
living cells, 43, 114
living matter, 44, 119
Locke, John, 66
LOGIC PUZZLE, 24
logical sequence, 22
logical thinking, 22
love, 67, 69, 70, 76, 77, 85, 120,
    126, 135, 136, 141, 142, 162, 165
lovers, 83, 90, 109, 115, 135, 137
Lucretius, 168-169
Luther, Martin, 164

M

Magee, Bryan, 66
Magic Numbers, 101
magnetic lens, 35
magnetism, 34, 58, 172
malady, 122, 156
mankind, 13, 14, 17, 18, 41, 45, 50,
    75, 94, 155, 163, 164, 169, 170,
    174, 175
marriage, 71, 90, 142, 150
mathematical probability., 18, 98
mathematicians, 13, 23, 47, 173, 174
mathematics, 16, 23, 28, 47, 57, 76,
    77, 121, 122, 139, 154, 163
Memes, 71
memory, 60, 62, 68, 94, 123, 124,
    150, 151, 153, 157
memory bank, 122, 152, 169
men-of-God, 14
men-of-science, 14
mental activity, 20, 88

mental pendulum, 81
mental picture, 18, 20, 27, 97, 127
metaphysical, 6, 121, 124
micro-waves, 57
Middle East, 139
mind, 6, 15, 17, 22, 26, 34, 36, 60,
    61, 62, 65, 67, 68, 74, 75, 77, 78,
    81, 82, 83, 84, 86, 88, 89, 90, 91,
    92, 109, 110, 111, 112, 117, 118,
    119, 121, 122, 124, 125, 131,
    133, 135, 136, 137, 140, 141,
    142, 143, 144, 147, 148, 149,
    150, 151, 152, 153, 156, 158,
    160, 162, 163, 165, 167, 168, 169
mind/soul link, 110, 111, 112
modern motor car, 93
modern telescopes, 35
molecules, 42, 97, 114
Monism, 13
Montaigne, Michel de, 145
Moon, 46, 92
Moses, 14
mother, 32, 62, 65, 67, 88, 133, 141,
    147, 156
music, 25, 53, 57, 58, 72, 76, 83, 84,
    85, 91, 94, 121, 135, 136, 139,
    146, 152, 153, 163
Muslim, 87, 117, 134, 139
mysticism, 13, 75
myth and legend, 17, 161, 162, 164,
    170, 173

N

natural occurrences, *25*
nature, 17, 19, 23, *24*, 27, 28, 35, *38*,
    47, 51, 53, 66, 72, 74, 77, 84, 90,
    98, 103, 116, 132, 140, 149, 157,
    159, 160, 161, 162, 164, 166,
    168, 171, 172

Neanderthal Man, 76
nervous net', 64
neurons, 58, 66, 89
neuroscientists, 7
neutrons, 40, 41, 42, 97, 99, 103,
    114, 123, 170
Newton, 14, 39, 94, 98, 162, 166,
    173
Nitrogen, 42, 43, 44, 100, 157, 170
non-Euclidean geometry, 27
Northern Ireland, 135, 139
noxious waves, 82
nucleus, 18, 40, 41, 42, 97, 98, 101,
    103, 104, 106, 127

O

objective principles, 38
open mind, 10, 13, 25, 112, 113, 174
operating system, 61
orbital radius, 98
organizing one's thoughts, 36
organize one's mind, 20
Origin of Species, 45
origins of life, 10
orphan, 67
out-of-phase, 55, 91, 173
Over, Raymond Van, 110
Oxford English Dictionary, 38
oxygen, 42, 43, 153, 157, 170
Oxygen, 44, 100

P

painting, 25, 83, 165
parental vibes, 87
parents, 30, 65, 68, 70, 74, 86, 91,
    95, 138, 141, 142, 143

particles, 10, 18, 30, 41, 44, 51, 98,
    101, 102, 103, 104, 106, 108,
    114, 127, 128, 161, 166, 170
perception, 27, 95, 97, 144
personal aura, 59
personal computer, 23
personality, 37, 66, 67, 69, 72, 81,
    83, 88, 108, 109, 110, 111, 125,
    131, 157, 158, 165, 173
personality code, 66, 67, 110
philosophers, 7
philosophical, 7, 34, 36, 94, 97, 126,
    128, 155
philosophy, 66, 107, 124, 152
philosophy of life, 25, 50
photon, 102
physical truth, 14
physical universe, 38
physicists, 13, 47, 114, 127, 161,
    166
physicists and mathematicians, 7
picture language, 61
Planck, Max, 101
planets, 38, 44, 67, 92, 98, 137, 162
Plato, 173
pornography, 149
positron, 103
precedence, 65
premonitions, 110, 111
primary waves, 91, 103
primitive life, 45, 49
primordial gases, 73
procreation, 60, 72, 149, 158
Protestant, 135, 139
protons, 40, 41, 42, 97, 99, 100, 103,
    114, 123, 170
Ptolemy, 38, 173
pure energy, 39, 40, 43, 45, 75, 97,
    98, 101, 102, 103, 104, 109, 111,

114, 115, 124, 125, 127, 128,
160, 162, 170, 171, 172
Pythagoras, 162, 163, 173

Q

Quantum Mechanics, 106
quantum of energy, 98
quantum physics, 6, 10, 18, 104
quantum theory, 101, 106
quarks, 18, 40, 103, 104, 114
quest for knowledge, 26

R

radiation, 29, 33, 39, 73, 78, 101,
102, 135, 137, 155, 156
radiations, 64, 73, 115, 116, 121,
125, 155, 172
radio-waves, 57
re-creation, 162, 172
reductionism, 10, 97, 161
relativity, 30, 31, 32, 33
religion, 10, 15, 16, 36, 95, 130, 132,
134, 147, 161, 164, 166, 171
religious doctrines, 13, 15, 38, 90,
160
religious faith, 38, 75
religious teachings, 15, 87, 162
resonance, 79, 80, 84, 117, 126, 135,
144, 153, 157
Rieu, Andre, 84
right and wrong, 10, 67
robots, 63
Rosetta Stone, 7
Russell, Bertrand, 87

S

sadistic vibrations, 80

sages, 90
savants, 91
Scandium, 100
Schopenhauer, Arthur, 69
science, 10, 13, 15, 16, 17, 18, 24,
28, 34, 36, 38, 39, 40, 76, 95, 96,
122, 132, 161, 164, 166, 171,
174, 175, 180
scientific proclamations, 14
sculpture, *25*, *83*
seers, 91
self, 19, 60, 67, 125, 126, 127, 163
sexual, 61, 69, 70, 72, 90, 149
sexual tendencies, 70
Shakespeare, 77, 131
siblings, 68, 90, 142
sideways thinking, 24
sine wave, 53
singularity, 104
Social Darwinism, 13
sociopathic behaviour, 81
Socrates, 19
solar system, 44, 97, 98
soul, 7, 10, 37, 67, 68, 73, 74, 84, 89,
109, 115, 117, 119, 120, 124,
125, 136, 137, 147, 156, 158,
163, 170, 172
sound waves, 33, 53, 57
soup of experiences, 9
space, 6, 26, 27, 29, 30, 33, 34, 35,
36, 39, 44, 58, 63, 73, 104, 119,
128, 136, 169, 170
space curvature, 34, 104
space probes, 15, 64
space travel, 33
space-time, 6, 10, 27, 73, 74, 119,
124
spatial cognition, 57
speed, 28, 29, 30, 32, 33, 34, 35, 53,
62, 63, 104, 110, 111

speed of light, 29, 30, 33, 34, 35, 39,
   62, 104, 110, 111
speed of sound, 33, 53
Spencer, Herbert, 13
spherical wave, 98, 101, 106
spiritual awareness, 18
spiritual world, 6
spirituality, 67
stardust, 44
stars, 12, 14, 38, 42, 44, 46, 73, 75,
   92, 170
structure of a concept, 23
structure of matter, 97, 107
sub-atomic, 6, 51, 103, 106, 108,
   123, 127, 157, 161, 166
sub-characteristic, 81, 110, 129, 131,
   149
sub-characteristics, 71, 109, 130,
   132, 157
subconscious, 22, 110, 141, 153
subversive signal, 80
Sun, 9, 12, 14, 16, 28, 44, 133, 164
superheated energy, 73
superimposed waves, 63
supernatural power, 19
supernova, 12, 42, 49, 114
supernovae, 64, 67, 73
synapses, 58, 89

T

technology, 57
terrorism, 87, 118
Thales of Miletus, 171-172
The Church, 13, 14, 171
The Creation, *38, 65, 67, 73, 75, 135*
the mind, 22, 23, 24, 26, 29, 32, 37,
   49, 59, 64, 66, 67, 71, 72, 73, 77,
   91, 92, 109, 110, 119, 125, 151,
   153

theologians, 7, 10, 162, 174
theory of everything', *9*
theory of evolution, 13
theory of gravitation, 39
theory of relativity., 30, 31, 33
think in pictures, 20, 77, 97
think logically, 22
think objectively, 20
think things through, 19, 20, 22
thinking, 10, 17, 19, 20, 24, 29, 34,
   35, 36, 46, 47, 59, 66, 68, 71, 72,
   75, 79, 83, 84, 88, 90, 93, 107,
   118, 119, 123, 154, 156, 158,
   162, 164, 165, 175
thinking brain, 8
thinking illogically, *22*
thinking person, 7, 72
thinking sideways, *22*
thinking-human, 94
thought process, 37, 61
thought processes, 8, 11, 19, 121
time, 26, 27, 28, 29, 30, 31, 32, 33,
   35, 36, 39, 46, 62, 66, 67, 69, 91,
   92, 94, 104, 109, 110, 114, 115,
   116, 119, 130, 132, 136, 148,
   154, 173
Torah, The, 14
Tree of Science, 17
true nature of science, 13
truth, 8, 13, 15, 16, 17, 19, 20, 25,
   36, 38, 48, 75, 91, 102, 119, 148,
   162, 164, 170, 171, 173, 174, 175
tsunamis, 73
tunnel vision, 10, *25*
Tylden, Mark, 63-64

U

ultraviolet waves, 58
Uncertainty Principle, 106

understanding others' viewpoints, *25*
unified theory, 7
uniform, 81
universal, 29, 57, 127, 128, 134, 161,
    162, 165, 167, 170, 171, 172, 173
universal link, 74
universe, 12, 14, 15, 16, 18, 38, 39,
    40, 41, 42, 43, 44, 46, 73, 76, 94,
    97, 102, 104, 108, 109, 114, 116,
    121, 125, 128, 135, 162, 164,
    169, 170, 171, 172, 174
untrained mind, 10

V

velocity, 31, 33, 35
vibes, 65, 68, 74, 83, 89, 90, 126,
    134, 136, 138, 139, 140, 141,
    142, 153
vibrating shell, 98
vibrations., 53, 70, 112, 117, 135,
    166
virtual-reality', *36*
vision, 9, 40, 77
Voltaire, 87

W

Wagner, 82
warped mind, 80, 82
warping of space, *35*
Watchtower Bible and Tract Society,
    45-47
Watson and Crick, 60, 94
wave/particle paradox, 98
waveform, 56, 65, 66, 88, 92, 118,
    119, 122
wavelength, 29, 33, 53, 54, 55, 57,
    69, 78, 80, 88, 90, 98, 100, 101,
    118, 122

waves, 29, 30, 33, 34, 35, 40, 41, 43,
    51, 53, 55, 56, 57, 58, 61, 62, 63,
    64, 65, 66, 67, 68, 72, 73, 74, 75,
    78, 80, 82, 83, 84, 87, 89, 90, 91,
    92, 93, 94, 97, 99, 101, 102, 103,
    104, 106, 107, 108, 109, 110,
    111, 113, 114, 115, 116, 117,
    118, 119, 120, 121, 122, 123,
    124, 126, 127, 128, 129, 131,
    135, 137, 138, 140, 144, 146,
    150, 152, 153, 155, 156, 158,
    160, 161, 162, 163, 164, 165,
    166, 167, 169, 170, 172, 174
Waves of terror, 80
Weinberg, Steven, 116
What actually is harmony?, 53
What are emotions?, 126
what control does He exercise over
    our lives?, 15
what have I learned from life?, 9
What is conscience?, 125
What is life all about?, 7
What is meant by *self*?, 125
What purpose has my life served?, 7
Wheaton College, 86-87
Wheeler, John, 28
Where do we go from here?, 7
Where is the soul?, 124
Whitehead, A.N., 160
Why am I who I am?, 7, 125
William of Conches, 15
will-to-life', 69
WW2, 94, 148

X

X-rays, 57, 58, 108

Z

Zimmer, Carl, 86

Printed in the United States
By Bookmasters

Printed in the United States
By Bookmasters